EcoBeaker 1.0

EcoBeaker 1.0

An Ecological Simulation Program

Eli Meir

University of Washington

 Sinauer Associates, Inc. • Publishers
Sunderland, Massachusetts U.S.A.

EcoBeaker 1.0: An Ecological Simulation Program

For information or to order, address:
Sinauer Associates, Inc.,
P.O. Box 407
Sunderland, Massachusetts 01375-0407 U.S.A.

FAX: 413-549-1118.
Internet: publish@ sinauer.com

ISBN 0-87893-523-1

 Printed on recycled paper with 10% post-consumer waste

Printed in U.S.A.
5 4 3 2 1

Table of Contents

EcoBeaker Laboratory Guide

EcoBeaker Manual

Included Procedures

Preface

EcoBeaker

Welcome to the world of EcoBeaker, a program that lets you design and play with a wide variety of ecological models. EcoBeaker is an ecological simulation program, designed primarily for use in the classroom although it can also be useful for researchers interested in quickly constructing ecological models. EcoBeaker gives you a two-dimensional world in the computer, upon which you can place creatures whose behaviors you design. You can then watch as the creatures eat, reproduce, move around, die, and do all the other things that creatures normally do, producing patterns that you can compare against the real world and against theoretical predictions. If you want to get more quantitative you can also make graphs of many different statistical measurements from the EcoBeaker world and sample the populations using a variety of common sampling techniques.

I have tried to make EcoBeaker easy to use at a variety of levels. For the student and teacher, I have prepared a set of laboratories demonstrating principles and examining specific problems in ecology and conservation biology. For the more intrepid student or teacher, I have designed EcoBeaker so that users can easily make their own models, by mixing and matching the included sets of behaviors to make unique species. For the programmers among you, or those of you looking for an individual-based modeling system for use in your research, I am including source code that lets you write your own behaviors for species in EcoBeaker models. See the next few pages for my specific notes to students, teachers, and researchers.

In addition to being easy to use, I have also tried to make EcoBeaker fun to play with. I have really enjoyed writing the program over the last several years, and I hope that this enjoyment and feeling of exploration is passed on to you as you use it. I especially hope that those of you using EcoBeaker to learn something about ecology will approach the laboratories contained here as puzzles or games, in the same spirit as a scientist approaches nature. I hope that the learning happens in spite of itself because you are actually doing something and are interested in the outcome. Although a computer program can never substitute for the real thing (much as the sit-in-a-chair scientists among us might try to ignore this fact), I hope that playing with EcoBeaker feels a little like exploring life itself.

A Note to Students

The first part of this book contains a series of laboratories that I've prepared to illustrate different ecological concepts. Perhaps laboratory is a bad name, because it conjures up images of damp, smelly rooms with black bench-tops, test-tubes all over, and students trying hard to look like they're not worried about the grade they're going to get. But I'm using laboratory in a good sense, to divide your explorations with EcoBeaker into bite-sized sections. I also try to include some background information with each lab, telling what grand old ecological principle it is supposed to illustrate. The spirit of EcoBeaker, though, is one of fiddling around, and I hope that's what will happen. I hope that whichever laboratories you end up doing, you'll start playing with them, changing them, and discovering things for yourself, or at least having fun seeing what appears on the screen. The numbers and exercises are intended to be just guides, to show you step by step some of the things that I've done and found interesting and to help get you started.

The second part of this book is a manual for EcoBeaker. It describes everything you can do with the program. You shouldn't ever need to look at the manual if all you are doing is following the instructions in one of the labs. If you get confused by something, however, or if you would like to go beyond what's in a lab (or design your own labs), then the manual should tell you what you need to know.

A Note to Teachers

Thanks for using EcoBeaker in your class. I have tried to make all of the materials as self-explanatory as possible, so with a bit of luck, using EcoBeaker shouldn't require inordinate effort on your part. This note contains a few thoughts based on my experience using EcoBeaker in classes at Hebrew University and the University of Washington, and on feedback I've gotten from all the students and teachers who have helped me over the last couple years.

This book is split into two parts. The first part is a series of laboratories, and the second part is a manual for the program. Each laboratory is supposed to illustrate one set of ecological concepts or involve the students in some ecology-oriented experiment. I have tried to make these laboratories self-contained. Students will not need to refer to the manual in order to use the labs, and for the most part, neither will you. In order to know what's going on a little better, though, you may want to skim through the "How EcoBeaker Works" section of the manual, and look over my descriptions of the procedures used in any laboratory you assign. You may also want to motivate each lab you assign, either by giving a lecture on the same general topic as the lab or through a short discussion before students begin work. These discussions seem to help increase student interest a lot and to enhance the written introductions at the beginning of each lab.

When we use EcoBeaker, we give students two options for their work location. Most students prefer to take a copy of the program with them to their home computers or to student computer centers on campus. In our experience, these students usually have no problems following the instructions in the lab guide and completing the assignments on their own. For those students who don't have easy access to a computer or who are not as comfortable using computers, we also make the program available in the Biology department computer lab, where there is someone to assist them. This latter setup has the additional advantage that students who have questions about the ecology in the labs can ask a teacher or TA. In some labs, it also might be nice to split the work between several groups of students and then have them present and combine their results at the end.

We have always had students work in groups of two to four (I've heard a rumor of a study that showed a group of three was optimal for computer work). We think that the interactions within a group of students promotes more learning than would occur if each student worked alone. Having students turn in group assignments also (incidentally, of course) makes the grading manageable.

I have tried to include laboratories here that cover a range of topics in ecology. Every lab included has been classroom tested and has received positive feedback from both students and teachers. I am greatly indebted to everyone who tried one of these labs, or one of the many other labs that I omitted in the end, and especially to those instructors who took the time to give me feedback and suggestions for improvement. I and others have written many other laboratories that didn't make it into this book. I am making these available on my web page (see My E-mail and Web Page below), so please "browse my page" to see if there is anything of interest to you. I am constantly developing new laboratories, so if you have a suggestion, I would like to hear it. I am also always looking for reviewers to try new labs for me. A list of the labs for which I want reviewers is given on my web page. If you are interested in any of them, please e-mail me. If you have any other suggestions, bug reports, or other comments, I would really appreciate hearing from you.

There is no reason for you to be restricted to the labs I supply, however. Quite a few instructors have modified one of my labs to better suit their purposes, or have constructed a lab of their own. To help you in modifying my labs, I have included a copy of this lab guide on the disk (see the "Special Installations" section). To modify the situation files or to make your own situation files, see the instructions in the second half of this book, the EcoBeaker Program Manual.

I have tried to design a lot of flexibility into EcoBeaker, and you should be able to construct a wide range of models from inside the program. Sometimes, though, you will think of a lab that can't be done with the procedures I've given. In that case, if you are a programmer (or can convince a programmer to help you), I have supplied enough source code for you to write your own modeling procedures and use them in EcoBeaker models. See my "Note to Researchers" for more information.

I hope that EcoBeaker proves useful in your classes and that both you and your students have fun with it. Again, I welcome any feedback you have, so I can make the next version even better.

A Note to Researchers

I did not design EcoBeaker to do research, but despite that, several people have found it useful for asking questions with individual-based models. If you want to try using EcoBeaker as the basis for publication-quality work, there are several things you should know.

First of all, I have included many sets of rules in the program (called Procedures) to govern how species behave in the models. I designed these procedures based on nothing except my faulty intuition about how things generally work (and I am no naturalist). I also endeavored to keep each procedure simple and with a minimal number of parameters so that it's easy to understand. While these procedures may be useful for you to explore and sketch out a model, you will almost certainly want to write your own, more precise, procedures for the model you use to get publishable results. I have tried to make writing an EcoBeaker procedure as painless as possible. On the disk, I include a programming manual and the source code you will need to write your own procedures. However, you'll need to download some more material from my web page as well (see the "Special Installations" section of the manual).

One thing you may want to change is my random number algorithm. I have used the Macintosh built-in random number generator, and am getting random numbers by selecting the low bits. I have read that this is a reasonable way to get random numbers on the Macintosh, despite the warning against this method in texts on numerical computing. I have not extensively checked it myself, though, since it's fast and works fine for teaching. In the source code I have included a slower but tried-and-true random number generator, and you will probably want to switch to that one in your models.

Another big caution is that, while I have fixed every bug I found in the modeling algorithms, there are undoubtedly still some left. It would be very unwise for you to spend a lot of time getting results from an EcoBeaker model before you thoroughly check to make sure it's running correctly. This advice goes doubly for procedures you write yourself—these rules for individual-based creatures can have some fairly subtle bugs that you won't notice unless you look closely.

If you do try to use EcoBeaker for research (successfully or not), I would be very interested in hearing about your experiences and your recommendations for any improvements I can make. One modification I hope to add to the next version is a macro language that will allow you to automate running an EcoBeaker model. I am also working on ways to make programming new procedures easier. In any case, if you do try EcoBeaker for a research project, I hope it proves useful in some capacity.

Enjoy!

My E-mail and Web Page

If you have bug reports, comments, suggestions, questions, or want to get in touch with me for any reason, please e-mail me at:

ecobeaker@zoology.washington.edu

Much of the program as you see it now has incorporated feedback from previous users, so I really welcome any suggestions you want to give me. I work on this program in my spare time, so I cannot test the program for bugs as thoroughly as I would like. Thus if you find any bugs or problems while using the program, please tell me. Bug reports are most useful if you can describe a series of actions that will reproducibly cause the bug. Thanks in advance.

I have many laboratories that, for one reason or another, didn't make it into this book. I am also developing new laboratories for the next edition of EcoBeaker. In addition, I will periodically release updates to EcoBeaker that contain bug fixes and new features. All of these are stored on my web page. To get to my web page, point your web browser to:

http://www.webcom.com/sinauer/ecobeaker.html

Acknowledgments

Thanks to everyone who reviewed the program for me and offered suggestions, especially John Banks, Tom Bannister, Ken Baxter, Timothy Bell, Paulette Bierzychudek, Daniel Brumbaugh, Ragan Callaway, Ted Case, Dean Cocking, Bill Fagan, Chuck Greene, Louis Gross, Drew Harvell, Jon Herron, Jon Jay, Francis Juanes, Junhyong Kim, Joel Kingsolver, Stacy Kiser, Frank Kuserk, Carol Lee, John McCarty, Gretchen Meyer, Daryl Moorhead, David Morrison, Garry Odell, Michael O'Donnell, Bob Paine, Ingrid Parker, Miguel Pascual, Bob Podolsky, George Robinson, Jennifer Ruesink, Raymond Russo, Neil Sabine, Biagio Sancetta, Cheryl Schultz, Ellie Steinberg, Mats Svensson, Dan Udovic, Peter Waser, Jordan West, and Arthur Woods (and anyone else who I have mistakenly forgotten to include).

Many thanks to Adrian Sun for discovering lots of bugs, making suggestions, and patiently listening to me moan. Also, thanks to Peter Kareiva for many useful comments.

Special thanks to Amatzia Genin and Emmanuel Noy-Meir of Hebrew University, who started me on this whole project and in whose class the program was first used, and to everyone at the H. Steinitz Marine Biology Laboratory in Eilat, Israel, where I wrote the initial version of the program. I'd especially like to thank Amatzia for letting me spend a good part of several months that I was supposed to be working for him on writing this program instead, and for generally being a great host while I was in Eilat.

Thanks to everyone at Sinauer Associates for their help in putting on the finishing touches.

Finally, thanks to all the students who suffered with earlier versions of EcoBeaker and told me exactly what they did and didn't like, and thanks in advance to all of you who are using the program now and who drop me a note to tell me what needs to get better.

Eli Meir

Eli Meir is currently supported by a Howard Hughes Medical Institute Predoctoral Fellowship.

To my parents, who taught me to think

EcoBeaker
Laboratory Guide

Oil Spills and Logistic Bacteria

Approximate time to complete: 1–2 hours

Background

One of the most basic ecological questions we can ask is, how fast does a population of some species grow? This is useful to know in all kinds of situations, from understanding how diseases spread to growing crops to saving endangered species. Growth rates are also an important part of more theoretical ecological questions. Because of this importance, it's worth having a good intuitive understanding of how populations grow. In this lab, you will take the role of a consultant for a company designing a bacteria to eat oil spills. The company wants you to tell them whether the bacteria they currently have grows fast enough to control oil spills, and if not, what they should do to try and make it better. As we do the experiments to answer these questions, you will hopefully get a feel for how a population grows in a simple environment, providing the basis for understanding more complicated situations.

3

Outline of This Lab

The world today runs in large part on oil, and this oil moves from one place to another on ships. All too often, one of these ships springs a leak, either by running into something, getting caught in bad weather, or through some other type of mishap (the captain got drunk at the wrong time and it goes downhill from there). These oil spills can cause havoc for the wildlife in the area of the spill, as is demonstrated most forcefully after a major spill by television images showing birds and other animals dying in the middle of the ocean because they are covered in oil and can't get it off. The oil can also contaminate beaches, and kill commercially valuable fish and other species. The effects of spilt oil can persist for years if the oil soaks into the mud, sand, and rocks near shore. Of course, not all spills are this bad—some spills happen in the middle of nowhere and the oil just evaporates or sinks to the bottom of the ocean. But enough spills happen in areas where they are dangerous, so that anything which we can do to lessen the impact of a spill is well worth exploring.

One new technology that shows some promise in getting rid of oil spills is spraying a spill with oil-eating bacteria. Although crude oil doesn't exactly look appetizing, it's filled with energy — that energy is what we use to run our cars and power plants. This energy can also be used by certain bacteria as a food source. If we could engineer an oil-eating bacterial species that grows really quickly, perhaps we could spray this bacteria onto oil spills and they would literally eat the spill away. Of course, one of the most important characteristics of this type of bacteria will be how it grows when there's lots of oil around, and that's what we'll explore here.

In this lab, you are a consultant to Tanker Toiletpaper, Inc., a firm specializing in cleaning up oil spills. They have just developed a new strain of bacteria for eating crude oil, and they want you to do some tests to tell them how close this bacterial strain is to being useful. To be useful and cost effective, they think that this bacteria must eat 90% of an oil spill within 2 days. Furthermore, to get this bacteria to the spill you need to fly a plane over the spill and drop the bacteria from the air. A small plane can't hold too much bacteria (relative to the size of a typical oil spill), so it's important that dropping just a small number of bacteria is enough to meet the 90%-gone-in-2-days criteria. Your job is to tell Tanker TP whether the strain of bacteria they have is good enough to do the job, and if not, whether they should invest time and money in making a better bacteria or invest in bigger airplanes that can carry more bacteria to a spill.

You have a laboratory setup that is very useful for doing these experiments. You can put a bit of seawater covered with a thin layer of oil onto specially designed slides, add a fixed amount of bacteria to the slide, and then put this onto a microscope. Your microscope is good enough that you can see individual bacteria. Furthermore, you have an instrument connected to the microscope that will count the number of bacteria on the slide for you and make graphs of this number over time. Using this setup, you should be able to help Tanker TP figure out what to do with their bacteria.

The Lab

1. Run EcoBeaker (double-click on its icon).

2. Open the situation Oil Spills (use the OPEN command in the File menu).

You should see several windows laid out on the screen as follows:

The slide on which we'll conduct experiments

A graph showing the amount of oil and bacteria over the course of each experiment

A table showing the same information as the graph above it, but as numbers

The window for changing the bacteria used in your experiments

The control panel used to control running and stopping the model

In the upper left is the Microscope Slide where we'll be doing our experiments. This slide is 1 mm on each side, and is marked off in a grid pattern, with each grid square big enough to hold about one bacterial cell. To the right are a graph and a table, which will show us how much bacteria and oil there is as the experiments progress. In the bottom is the Control Panel, used to control the experiments. Under the Microscope Slide is a window titled Experimental Parameters which we'll use later on to change the experimental parameters.

3. Start the experiment by pressing the GO button in the Control Panel. Wait until all the action finishes and then stop the experiment (press STOP in the Control Panel).

You'll see oil get spilled onto the water at time 0. Normally, it would take several hours after a spill for a crew to get to the spill with the bacteria, so to simulate this lag we don't add the bacteria until 6 hours after the experiment starts. Tanker TP has told you that it is economically feasible to add two bacterial cells per square millimeter, so we start out by adding only two cells to the slide (which, as you'll recall, has a size of 1 mm²). As the experiment runs, you should see bacteria moving around on the slide, eating the oil, and reproducing. When the oil runs out, you'll see the bacteria start dying from starvation.

You can look at the Population Graph to get a better idea of how the bacteria population grows and shrinks. If the interesting part of this graph has moved off the edge, you can scroll backwards using the scroll bar at the bottom of the graph. The scale at the bottom of the graph is in hours. Your experimental system also makes an exact count of the number of bacteria every 12 hours, and this number is shown in the Population Table.

4. Write down a short description of the growth of the bacteria, both from what you remember while watching the experiment and from looking at the graph.

5. Remember that the criteria for whether this bacteria was successful was that it should eat 90% of the oil within 2 days (48 hours) of the time of the spill. Using only two bacteria per mm², are they successful at cleaning up the spill? If not, how far off are they? How many more hours did it take? You can see the exact amount of oil left after each 12 hours in the Population Table. The initial amount of oil is about 2000 drops.

6. Since what happens is partially governed by chance, you should run the experiment at least a couple more times to see if your results from the first experiment were normal. To run the experiment again, reset it (click on RESET on the Control Panel) and then start it again (GO). For each run, figure out the length of time it took until 90% of the oil was eaten. Then average these results.

As you saw, the bacteria are not eating fast enough to do the job. Tanker TP wants to know both how much more of this bacterial strain they would need to use per area of oil spill and, alternatively, how much better a new strain of bacteria would need to be to do the job with only two bacteria per mm². In order to find the answers, we need to figure out exactly how fast the bacteria are growing. We can calculate the growth rate from the bacterial counts in the Population Table, which shows the number of bacteria on the slide every 12 hours from the start to the end of the experiment.

The growth rate of the bacteria is the number of new bacteria produced by each already existing bacteria in a given amount of time. Since we are measuring bacterial counts every 12 hours, it's easiest to calculate the bacterial growth rate per 12-hour period. A growth rate of 0.5 per 12 hours would mean that on average, each bacteria made one-half of a new bacteria in 12 hours (or to put it another way, one of every two bacteria reproduced in that 12-hour period). This growth rate then tells you how much the total population size of the bacteria went up over those 12 hours. If each bacteria is producing, on average, 0.5 new bacteria over the 12 hours, then the total population size will increase by 50%. We can also go backwards—if the population increased by 50% over the 12-hour period, then on average each bacteria must have produced 0.5 new bacteria, and so the growth rate must be 0.5.

7. Before calculating growth rates, try to predict when the bacterial growth rate will be highest. Are the bacteria growing fastest when they are first added to the oil, when their population size is highest, or somewhere in-between those two? Write down your prediction.

8. Calculate the growth rate of the bacteria. Remember, the growth rate is the number of new bacteria produced over the 12 hours by each individual bacteria which was already there. Calculate at least a couple growth rates near the beginning of the experiment, a couple when the bacteria are nearing their maximum population size, and one when the bacteria population is declining. Do the bacteria grow faster, slower, or at the same rate through the course of the experiment? Write down your answers, and a short explanation of why you think any changes in growth rate might be happening.

9. As above, you should repeat this experiment at least twice more to get average results (RESET the experiment and then run it again). To save time, when you repeat the experiment you can calculate growth rates just where they are going to be highest (as you determined in step 8).

Now we have the information that we need to start making recommendations to Tanker TP. First of all, let's figure out how much extra bacteria they would need to add in order to be pretty sure of meeting the success criteria. They want to add two bacteria per square millimeter of ocean, as we've done in the experiments up until now, but if they needed to add just one or two extra bacteria, then maybe that would still be cost-effective. We can figure out the effect of adding extra bacteria from the growth rate.

To use the growth rates you calculated, you would put them into a formula like the following:

$$N_t = N_{t-12} + rN_{t-12}$$

where N_t is the number of bacteria at time t, N_{t-12} is how many bacteria there were 12 hours before time t, and r is the growth rate. What this equation says, in words, is that the number of bacteria at time t is equal to the number 12 hours ago, plus the growth rate that you calculated above times the number 12 hours ago. You should be able to derive this formula from the formula you used to calculate the growth rate.

10. How will doubling the amount of bacteria we add initially (from 2 to 4) affect the time it takes them to eat 90% of the oil? It should take less time, but how much less? You should be able to calculate this from the growth rate that you figured out above. Write down how long you calculated it will take with the extra bacteria, along with the reasoning you used to make this calculation. If you are having trouble figuring this out, think about how long it took in the first experiment for the bacteria population to grow from 2 to 4 (which it no longer has to do when it starts at 4).

Now we'll see if this prediction is right. We can redo the experiments we did above with four bacteria added initially instead of two. Here's how to do that.

11. Find the Experimental Parameters window. The top item in this window shows how many bacteria to add to the slide at the beginning of the experiment. Right now it is set at 2. Replace the 2 with a 4, and then click on the CHANGE button at the bottom of the window.

12. Reset the model (RESET in the Control Panel), and run it again (GO). How long did it take for 90% of the oil to get eaten now? Repeat the experiment a couple times to make sure your results aren't just due to chance. Was this approximately what you predicted?

13. Now make a prediction for how long it will take to eat 90% of the oil if you double the starting number of bacteria again from 4 to 8. Write down this prediction. Then repeat steps 11–12 using 8 bacteria.

14. Is this a very efficient way of reducing the time it takes to eat the oil? Remember that each time you double the number of initial bacteria, you are also doubling the cost and effort of using the bacteria. So do you get much benefit by paying for twice as much bacteria?

    Instead of using more bacteria, Tanker TP can try to make a strain that is a bit more efficient at using the oil to grow. This would make the growth rate go up, because each of the bacteria wouldn't have to eat as much oil before reproducing.

15. Before we actually try increasing the efficiency of the bacteria, let's again try to predict what the result should be. If the bacteria are twice as efficient, this means they can reproduce twice as many times after eating a given number of oil drops, so that their growth rate should approximately double. From this, you should be able to make a rough calculation of how long it will now take for the bacteria to eat 90% of the oil, starting from just two bacteria. Write this prediction down, along with the reasoning you used to get to it. You will not be able to make this calculation exactly, so you don't neccesarily need to use the growth equation given above, but you should be able to get a pretty good rough answer just from thinking about what will happen if the bacteria reproduce twice as fast.

    The amount of energy that a bacteria gets from eating one drop of oil is shown as the second number in the Experimental Parameters window. Increasing this number will make the bacteria extract more energy from each drop of oil. This number is currently set to 2. Since we doubled the number of bacteria to see what effect that would have, let's also try doubling the efficiency. In order to make a fair comparison, we will first decrease the initial number of bacteria back to 2.

16. Reduce the starting number of bacteria back down to 2.

17. Increase the energy gained from a drop of oil from 2 to 4 by finding this item in the Experimental Parameters window, changing the 2 to a 4, and then clicking on the CHANGE button.

18. Reset the model and then run it. How long did it take for the bacteria to eat the oil this time? Was this what you predicted? Run the model a couple more times so that you can get an average result.

19. If both increasing the efficiency of the bacteria and using more bacteria cost about the same amount, would your recommendation to Tanker TP be to stick with the bacteria they have now and try to use more of it, or to attempt to develop a more efficient bacteria? Why?

Notes and Comments

This lab should have given you a feel for exponential growth, and how even a small number of individuals can quickly grow into very many individuals. The speed at which the population grew depended on two things. The first was their *intrinsic growth rate*. This is the bacteria's maximum reproduction rate—the speed at which they reproduced when they were completely surrounded by food. More generally, the intrinsic growth rate of a species describes how fast individuals can reproduce under ideal conditions for their growth. The only limits to growth of the population are then internal factors, such as how fast the bacteria can eat, how fast they can make babies, and so on. Early on, in each of the experiments we did here, the bacteria were growing at a speed quite close to their intrinsic growth rate.

Once the oil started to be eaten, the growth of the bacteria was limited by the amount of food available. Eventually, the bacteria ran out of food, and without food they died off. In lots of cases, this is the way that nature works—some new source of food appears, creatures rush to take advantage of it, use it up, and then all die off or wander away to find another supply of food. Dead plants and animals can work this way, as can animal waste products. There is a whole community of bugs in the African Savannah which takes advantage of elephant excrement, a lifestyle almost as appealing as eating crude oil. In many other situations, however, there is a steady supply of food instead of one burst of food. A given supply rate of food can support a certain number of individuals, and this number is called the *carrying capacity* of the environment. More generally, the carrying capacity is a measure of how large a population of the species can survive on a certain amount of resources.

When you changed the energy a bacteria gained from a drop of oil, you were changing both the intrinsic growth rate and the carrying capacity for the bacteria. You measured the increase in growth rate in the lab. The increase in carrying capacity wasn't directly shown, but you probably noticed that the peak bacteria population was higher when they were more efficient at eating oil. This increase in the number of bacteria came about because now the same amount of oil could support more bacteria, increasing the bacterial carrying capacity of the slide of oil.

Both the growth rate and the amount of resources it takes to support a given size population are routinely measured in many different situations. For instance, people who breed new crops want to know how fast their new strains grow, and how much fertilizer they have to use in order to grow a certain amount of the crop. Conservation biologists worried about an endangered species are interested both in how many individuals can be supported on the remaining habitat, and how quickly the remaining population could grow if more habitat was added. At the other end of the scale, people who study human demographics find the growth rates of human populations, use these growth rates to calculate how many people we'll have in the future, and argue over whether the earth's carrying capacity for humans is large enough to support these gigantic populations. Almost anywhere else you look in ecology, evolutionary biology, and related fields, you'll see growth rates and carrying capacities somewhere in the background. The types of observations that you made in this lab enhance our intuition in all these situations.

Competitive Exclusion Principle

Approximate time to complete: 1–2 hours

Background

The *competitive exclusion principle* is a theory that states that if two species are trying to occupy exactly the same niche, one of them will go extinct. In other words, no two species living in the same place can be exactly the same—if they are, one of them won't survive. In this lab we will test this theory by putting several species into a small world, each with exactly the same behaviors and eating the same food. If the competitive exclusion principle is correct, all the species except one should eventually go extinct. Your mission, should you choose to accept it, will be to try to prove the competitive exclusion principle wrong, or failing that, to figure out how different two species in this model world need to be to survive together.

Outline of This Lab

In this lab we'll do a simple test of the competitive exclusion principle. We'll fence off a big yard and put four different species of rabbits into it. Each species of rabbit is a different color, but in all other ways the rabbits start out the same as each other. Every day, we'll drop in a few leaves of lettuce for the rabbits to eat. During each day, every rabbit looks around itself for some lettuce to eat, and if it sees some, it hops towards it as fast as it can. When the rabbit eats a leaf of lettuce, it gains energy from it. The rabbit uses up energy each day just to keep itself alive, and if it can't get enough food to cover the energy it is using, it dies. However, if the rabbit manages to eat more than it needs just to live from day to day, this energy is stored. When the rabbit stores up enough energy it has a baby rabbit.

Initially, all the characteristics of each rabbit species are set exactly the same. These characteristics include how fast the rabbits can hop, how far a rabbit can see, how much energy a rabbit uses each day just to stay alive, and how much energy a rabbit accumulates before reproducing. You will be able to change each of these characteristics separately for each of the rabbit species, and you will also be able to change the amount of new lettuce that you are dropping in every day.

Because the rabbits are living together, eating the same food, and are identical in other ways, the competitive exclusion principle predicts that only one of these rabbit species can survive. To see if this is true, we'll let the rabbits go, then wait for a while and see what happens. I'll just remove a bit of the suspense right now by telling you that the way I initially set up the model, there is indeed competitive exclusion. Your job will then be to change the rabbits, the lettuce, or the yard somehow so that more than one rabbit species can coexist.

The Lab

1. Run EcoBeaker (double-click on its icon).

2. Open the situation Competitive Exclusion (use the OPEN command in the File menu).

You should see several windows laid out on the screen as follows:

Bird's eye view of the yard in which we'll be doing experiments

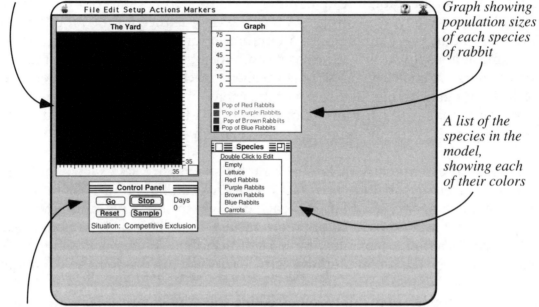

Graph showing population sizes of each species of rabbit

A list of the species in the model, showing each of their colors

The control panel used to control running and stopping the model

The window in the top left shows a bird's eye view of the fenced yard where we are conducting these experiments. The yard is 35 meters on a side. When we start the model running, you'll see grass growing in here and rabbits hopping around. To the right of the yard is a window with a graph showing the population size of each species of rabbit. Below this graph is the Species window with a list of all the species in the model. You'll click on the species in this window in order to change their parameters. Finally, below The Yard is the Control Panel, which has the controls for starting and stopping the model.

3. Run the model to get an idea of what's going on (push the GO button on the Control Panel).

You will see The Yard start filling up with light green lettuce. Pretty soon little colored bunny rabbits will appear and start hopping around eating the lettuce. As I explained above, each color of rabbit is a different species, but the rabbits are all exactly the same other than their color. In the upper right of the screen is a graph that shows the population size of each species of rabbit.

4. Watch the yard and the population graph for a while. Are all the species surviving? Keep watching. Does more than one species survive after you've watched for a while?

> I am going to take you through the motions of changing the parameters in the model so you can see how it's done. Let's start by changing how many lettuce leaves we throw into the yard every day. This number is governed by something that EcoBeaker calls a Settlement Procedure, which is just a set of rules saying how new creatures get into the yard. The settlement procedure of the Lettuce species is a set of rules called Fixed Settlement. Fixed Settlement says that every day we throw a fixed number of new lettuce leaves into the yard. Steps 5 and 6 say how to change that number.

5. Find the window entitled Species, and look in there for Lettuce. Double-click on LETTUCE (point the mouse at it and click the button twice in a row). A dialog box will appear where you can set up the rules governing the Lettuce species. Find the button labeled SETTLEMENT PARAMS and click on it. A second dialog box will appear that has the parameters for Fixed Settlement. You will see that the only parameter is the Number of Immigrants/Turn, which is set to 5, meaning that we're adding five leaves each day, thrown randomly around the yard. To change this to ten leaves per day, replace the 5 with a 10 and then click OK in both dialog boxes.

6. Now reset the model (push the RESET button in the Control Panel). Then run the model again (GO). Again watch The Yard to see how long more than one species of rabbit survives. Does anything happen differently with more lettuce?

> That's all there is to the lettuce. Now let's try changing one of the species of rabbits. Each species of rabbit has a settlement procedure similar to that of the lettuce, which determines how many rabbits from that species are dropped into the yard. The only difference is that we only add rabbits once, unlike the lettuce, which is added every day. Rabbits also have a second set of rules that tell what they do once they land in the yard. EcoBeaker calls this second set of rules an Action Procedure, and the action procedure of the rabbits is called Predator. Predator makes the rabbits behave as was described above in the outline of this lab. While rabbits are not normally thought of as predators, if you look at it from the perspective of the lettuce they might as well be predators, so EcoBeaker's set of rules for predators works fine for rabbits eating lettuce.

7. Double-click on the species RED RABBITS in the Species window to bring up the dialog box where you can set the rules for red rabbits.

   Clicking on the SETTLEMENT PARAMS button will let you change the number of red rabbits that are initially dropped into the world. Changing this number works the same way it did for the lettuce (click on the SETTLEMENT PARAMS button to change the number of red rabbits that settle). In the rabbits' settlement procedure there are two numbers, one saying how many rabbits to add and the second saying what day to add them. Initially the model is set up so that ten rabbits of each color are added on day 10. Waiting a few days before adding the rabbits means there is some food around for the rabbits to eat when they arrive.

8. To change how the red rabbits behave once they're in the yard, find the Action Params button and click on it. A second dialog box will appear with a few numbers and buttons. This is where you can modify the behavior of the red rabbits.

   Here's an explanation of all the parameters you see. Remember that every day, each rabbit looks around for food, and if it sees some, it hops towards it as fast as its little bunny feet will go. The rabbits can see Distance to Look squares away, and can hop Speed squares per day. If the rabbit doesn't see any food, then it will just hop around randomly. Hopping around costs the rabbit some energy, which is given by Cost of Living. When a rabbit lands on a square with food, it eats the food and gains the amount of energy specified in Prey Value. You can see that the Prey Value is currently set to 4, so if a rabbit eats a leaf of lettuce then it can live for four days on that meal (since the Cost of Living per day is 1).

   The species that a red rabbit considers food is given in Prey Species. Clicking on the Prey Species button brings up another dialog box with all the species in the model. The red rabbits will go after and eat any species whose check box is checked. Currently, only the Lettuce is checked, meaning that red rabbits will only try to eat lettuce, and don't consider any other rabbits as acceptable food. To turn red rabbits into predators on other rabbits (or cannibalists on themselves), you would check the boxes next to the rabbit species you want them to eat. Similarly, if you wanted them to stop eating lettuce, you would uncheck the box next to Lettuce.

   If a rabbit manages to get more food then it needs just to live, it stores up the extra energy, and when it gets up to the amount of energy indicated by Energy To Reproduce, then it makes a baby rabbit. These are kind of odd rabbits—they don't have sex, and when they have babies, they just split in half, like some bacteria. So the energy of the rabbit is split as well, half going to the baby and half going to the old rabbit. If any rabbit ever goes to an energy level of 0, it dies.

The last rule is that the first colonizer rabbits (the ones added by the settlement procedure) start with an energy level equal to half of Energy To Reproduce.

9. Let's say we wanted to make red rabbits faster and wanted them to be able to go longer between meals. Find the Speed parameter and change it from 1 to 2, doubling the speed that red rabbits can hop. Then find the Cost Of Living parameter and change it from 1 to 0.5, so the rabbits spend half as much energy hopping around each day. Then click OK in both dialog boxes. Note that you have changed the Speed and Cost Of Living for red rabbits only, not for the other species of rabbits. You can change parameters for any of the other rabbit species in exactly the same way.

10. It's not a bad idea to try changing only one thing at a time, so you may want to change the rate at which lettuce is added into the yard back to its original value (see step 5).

11. Now reset the model (RESET) and run it (GO). Does more than one rabbit species survive when red rabbits are stronger and faster than the other species? If not, is anything else different about the way things go? Write down your observations.

Here's a summary of the parameters in the model that you may want to change.

For the Lettuce species

Num Immigrants/Turn	The number of lettuce leaves you throw into the yard each day.

For each rabbit species

Num Immigrants	The number of rabbits of this species that are initially added into the yard.
Time Step to Settle	The time (in days) at which you add the rabbits of this species into the yard.
Distance to Look	The distance (in squares) that a rabbit can see when looking for food.
Speed	The number of squares a rabbit can hop each day.

Cost of Living	The amount of energy a rabbit uses per day just to stay alive.
Prey Value	The amount of energy a rabbit gains each time it eats some food.
Prey Species	A list of the species that a rabbit will eat.
Energy to Reproduce	The number of energy units a rabbit will build up before reproducing.

12. All right, now it's up to you. You can change any of the parameters in the model, and your goal is to try to get more than one species of rabbit to survive. Be careful: changing some parameters might lengthen the time it takes for species that are headed to extinction to actually go extinct, so don't declare victory too early. You'll have to decide for yourself when is too early.

    <Optional> You may have noticed that in the model there is a sixth species called Carrots. This is another food item that you can add into the yard. The Carrot species works exactly the same way as lettuce, but I initially set the number of carrots getting thrown into the yard every day to 0, so that there weren't any carrots. You can add in carrots and see if more than one species of rabbit can coexist when there are two different food sources.

13. To start throwing carrots into the yard along with the lettuce, double-click on the CARROTS species in the Species window, bringing up the species dialog box. Next click on the SETTLEMENT PARAMS button. You will see that the Num Immigrants/Turn is currently set to 0. Change this to the number of carrots you want landing in the yard per day, and then exit the dialog boxes.

14. You will also need to change some of the rabbit species so that they eat carrots instead of lettuce, or both carrots and lettuce. You do this by changing the Prey Species of those rabbit species, as described above under step 8.

15. Now see if with two different food sources you can get two species of rabbit to coexist. If you do, is this refuting the theory of competitive exclusion?

16. Once you have two species coexisting, see if you can get three species to coexist. Can you come up with a hypothesis for the number of species that can coexist on a given number of food sources?

More Things to Try

When you've convinced yourself that competitive exclusion is a good or bad theory for this model, you might want to try a few other things. If you have done the optional bit of the lab and now have two types of food items, you might want to make all species of rabbit eat both food items, but have carrots and lettuce be worth different amounts of energy. You can do this by changing the Action Procedure of the rabbits from Predator to Picky Predator. To do this, go into the species dialog box for one of the rabbit species. Find where it says Predator. Click and hold down the mouse button on PREDATOR. A menu will pop up giving a whole bunch of different sets of potential rules, including one called Picky Predator. Select PICKY PREDATOR and let go of the mouse button. Now you can change the parameters for Picky Predator just as you did above for Predator. All the parameters are the same, except that, instead of a Prey Species button, there is now a Prey Values button. When you click on this button, a dialog box will appear where you can give different prey values for each type of food. There is also one more new item called Time to Switch which you can read about in the manual.

Another variation to try is adding in disturbance to the model, and seeing what that does. You can add in hurricanes, which come by every now and then and randomly kill some of the rabbits. See the "Hurricane Settlement Procedure" in the manual, along with the description there of how to add species. You could also try dropping the Lettuce in patches instead of uniformly across the yard. To do this, make a set of habitats and use the Fixed Habitat settlement procedure to make the lettuce to land only in one habitat. Again, consult the manual on how to do this. You might also want to change the size of the yard in which you're conducting the experiments. To do this, select the GRID item from the SETUP menu, and look in the top of the dialog box that appears for two inputs called Grid Width and Grid Height. These give the width and height of the yard that you are using; changing them will change the size of the yard. For most of these alterations, you'll probably want to increase the amount of grass and spinach being thrown into the yard, so that the population sizes of the rabbits will be higher.

Intermediate Disturbance Hypothesis

Approximate time to complete: 1–2 hours

Background

Most habitats are subject to disturbances. A disturbance can be anything which changes that habitat, usually fairly dramatically. Forests, for instance, are subject to fires and blow-downs. Plants living on sides of hills or mountains get hit by landslides. Similarly, marine systems such as coral reefs can be decimated by storms, ravaged by predators, or trashed by inconsiderate tourists.

Disturbances can change which species live in a habitat, and can also change how common each species is. In general, when there is very little disturbance in some habitat, then the species that are best at competing with other species over the long term will eventually take over. When there is a very high level of disturbance, then species that can recover from a disturbance quickly or colonizing species that can move into a disturbed area quickly will take over. Thus, as the size or frequency of disturbance in an area changes, you will see changes in the distribution of species in that area. For instance, in a forest that has been around for a long time without being burned, you find certain species of trees and other plants that can grow under low light, but you don't find many grasses or bushes. If the same area were burned periodically, then those trees would never get established. Instead, you would get a prairie with grasses and other plants that could quickly move back and grow up fast following a fire.

There is a theory called the *intermediate disturbance hypothesis* that tries to predict how species diversity will change with changing levels of disturbance. Species diversity is some measure of how many species there are in an area, and how rare or common each species is. A rain forest has a very high species diversity, a temperate forest has not as high species diversity, and the Antarctic has very low species diversity. The intermediate disturbance hypothesis states that intermediate levels of disturbance will lead to the greatest species diversity. This is because, as described above, in a rarely disturbed environment, the most competitively dominant species will take over, and in a very highly disturbed environment, only species that deal with the disturbance well will survive. In the middle, both types of species will be around, thus giving higher species diversity. Or at least that's what this theory would lead you to believe. In this lab we'll check whether this hypothesis really works in a simple model.

Before starting on the lab, we need to decide how we're going to measure species diversity. One way would be to look at the total number of species represented in the plot of land we're studying. However, if there are 9991 individuals of one species in the area, and 1 individual each of 9 other species, this is obviously not as diverse as if there are 1000 individuals of each species. Several statistics for measuring diversity have been devised that take into account both the numbers of species and the relative population sizes of each species. We'll use one of these here to measure the diversity in the model as we change the level of disturbance.

The diversity index we will look at is called *Simpson's index of diversity*. This index takes as input the population size of each species in the study area, and spits out a single number indicating diversity. This number will be low when diversity is low (it has a minimum of 1), and will get higher as diversity gets higher.

<Optional> For those of you who are mathematically inclined and interested in how the index is calculated, here is the formula:

$$\text{Simpson's diversity index} \quad D = \frac{1}{\sum_{i=1}^{S} p_i^2}$$

where S is the number of species, and p_i is the population size of species i divided by the total population size of all species (the proportion of individuals that are of species i). Note that if you look in different sources, Simpson's index may be defined differently. You should be able to find explanations of this index and others in any ecology text.

Outline of This Lab

In this lab, we will look at a forest ecosystem in the eastern part of the United States that is being ravaged by fires. Each time a fire comes through, everything in its path is burned to the ground. Out of this empty earth, first grass and other annual plants spring forth. These are followed by blackberry bushes and other shrubs, then by white pines, sugar maples, oak trees, and finally hickory trees, which are the *climax species*. If there were no fires, eventually this whole forest would be composed entirely of hickory trees. However, whenever a fire burns an area of the forest, the whole process starts again from scratch.

What we will do is vary the numbers of fires and how big each fire gets once it starts. This will give us different levels of disturbance of the landscape. For each level of disturbance, we will measure the diversity of species in the forest. From this data, we should be able to say whether the intermediate disturbance hypothesis seems to hold for this model.

<Optional> If you are interested in how the model works, here is a short description. This model is built around a *transition matrix*—a list of probabilities that specifies the chance of an individual from one group becoming an individual from another group in the next time step. In this case, the time steps are years. For instance, every year there is a chance that a patch of ground that had grass growing will now have a blackberry bush. Then the next year, there is a chance that the blackberry bush will be replaced by another shrub, a pine tree, or a maple tree. There is also a chance that the blackberry bush will remain there (no change in what's growing in that piece of land). Each of these chances is gathered together into something called a matrix, which is just a convenient way of organizing all these numbers. Every year, the program goes through each square of land, looks at what's there currently, then randomly determines what will be there next year according to the probabilities that the plant now occupying the square will be replaced by some other plant.

If you want to see how this works in practice, you should read about transition matrices in the manual, and then look at the species setup box for each of the species. To get to the species setup box of a species, find the name of the species in the Species window, and double-click on it. A dialog box will appear, and on the right of the dialog box will be the chance per year that a plant of this species will be replaced by a plant of each of the other species. Note that empty spaces also have a transition matrix. The transition matrix of Empty gives the chance for each of the other plant species to start growing in an empty square (this is colonization).

The Lab

1. Run EcoBeaker (double-click on its icon).

2. Open the situation "Intermediate Disturbance Hyp" (use the OPEN command in the File menu).

You should see several windows laid out on the screen as follows:

The plot of forest that we're experimenting with

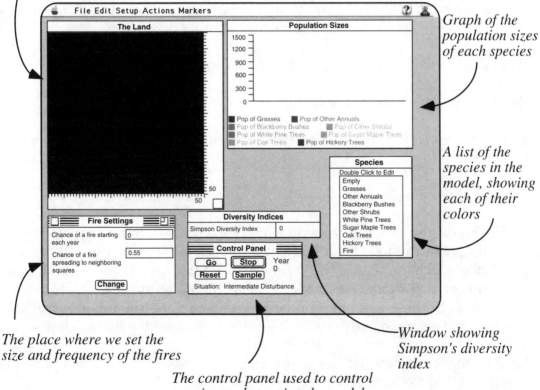

Graph of the population sizes of each species

A list of the species in the model, showing each of their colors

The place where we set the size and frequency of the fires

The control panel used to control running and stopping the model

Window showing Simpson's diversity index

The upper-left window shows the plot of forest that we'll be experimenting with. Black indicates land that has just been burned and so is empty of all species. Each plant species has its own color, and these are shown in the Species window in the bottom right. As the model is running, you can see how many individuals of each species are in the forest by looking at the graph of population sizes. Below that is a window that shows the value of Simpson's diversity index, which I discussed in the Background section of this lab. The window at the middle bottom is the Control Panel, which we'll use to control running the simulation, and the final window in the bottom left, Fire Settings, is where we can change the size and frequency of fires.

As you can see in the window showing the forest, the model starts out just after a gigantic fire burned down everything, so that the whole plot of land is empty (black indicates scorched earth). We'll start out by running the model without any fires. What do you think will happen? After the model has run for a while, how diverse do you think the forest will be? (Think about what will happen without fires according to the outline above.) What will the diversity index read (remembering that 1 means a single species and higher numbers mean higher diversity)?

3. Write down your answers to the above questions.

4. Run the simulation (push the GO button in the Control Panel).

5. When the forest has reached an equilibrium (nothing is really changing much), then stop the simulation (push the STOP button in the Control Panel).

6. How good were your predictions?

The next step is to start some fires. Look for a window labeled "Fire Settings" in the lower-left corner of the screen. There you will see two numbers, one the chance that a fire will start in a given year, and the other a number that determines how far on average the fire will spread once it starts. Both these numbers run between 0 and 1. You will notice that currently the chance of a fire starting is set to 0, so there are no fires initially.

7. Set the chance of a fire starting to something between 0 and 1. Then push the CHANGE button to change the settings in the model.

8. Reset the model so that we start from an empty patch of land again (press the RESET button).

9. Run the simulation again (GO).

Now watch what happens in the simulation. You will see little fires cropping up every now and then in different places around the forest, burning down everything in their path. This creates a patchy landscape, with different areas at different stages of recovering from burns, and a few areas that haven't been burned in a while. Patchiness is a big concept in ecology right now, with lots of work being done on whether and when it is significant (see Notes and Comments below).

10. Watch the diversity index until it seems to be more or less stable, or at least hovering around a certain value. Then stop the model, write down the level of disturbance (the chance of a fire starting and the chance of a fire spreading) and the value of the diversity index.

OK, now you have all the tools to test the intermediate disturbance hypothesis. Try out a few different settings for the fires to get a feel for the range of patterns you can get (remembering to push the CHANGE button each time you change settings). Then plan out a series of values that you will try with each of those parameters, to increase from low levels of disturbance, to moderate levels of disturbance, to high levels of disturbance.

11. Write down your planned experiments, and what you predict the diversity index will be at each disturbance level (or better yet, draw a predicted graph of diversity versus disturbance level).

12. Perform the experiments you outlined above. At each level of disturbance, write down the values of the parameters and the values of the diversity index.

Note that you are going to have to decide when to measure diversity, since it will go up and down as the fires come and go. So you have to pick some consistent measure of diversity—perhaps averaging it for a few years, or picking a certain year and always measuring in that year, or some other method. You might also note if the diversity varies more at some levels of disturbance than at others.

13. When you have several measurements of diversity at different levels of disturbance, make a graph of diversity versus disturbance level. What level of disturbance gives the most diversity? What level gives the least? Does your data support the intermediate disturbance hypothesis?

More Things to Try

If you want to play further with this model, here are some suggestions. Try playing with the transition matrices, and see both how they affect diversity and how they change the effects of the fires. See my description of transition matrices in the Outline section of this lab or in the manual. For instance, you could speed up the rate of succession by increasing the chance that each species will be replaced by another species higher up in the successional ladder. You could also try adding death to the simulation, by including a small chance that a square that holds a given species this year will be empty next year (set the chance of transition to Empty to be something larger than 0). You could also change the succession so it's not just a one way street. In the original model, grass can be replaced by bushes, which can be replaced by trees, but not the other way around (unless a fire comes and clears things out). Forests may indeed work this way, but in other ecosystems this sequence wouldn't be true. You could try adding in chances for the succession to go backwards. With all of these changes, take a look at the effects on diversity if there are no fires, and then run fires at a few different settings, as you did in the lab, to see whether the maximum diversity is achieved at the same setting as before.

Notes and Comments

Managers for park lands and preserves regularly deal with issues such as those we explored here. For years, there was a policy in many parks to try to put out fires as soon as they started, and not to let any forest burn. More recently, managers have started considering fires and other disturbances as part of the natural processes that help to rejuvenate the land and perhaps lead to a greater diversity of life. This effect may or may not be due to what we saw in this model—that having some disturbance (but not too much) lets in colonizers that normally would be out competed by other species. Certainly, there are other factors as well. For instance, the seeds of some species of trees will germinate only after a fire has come through. However, regardless of whether the intermediate disturbance hypothesis holds, there is no doubt that disturbance is a major factor in structuring ecological communities.

Another effect you saw in this lab was a patchy distribution of species, or *patchiness* for short. Patchiness means that when you look closely at a chunk of land that might look fairly uniform from afar, you will see that some parts have different collections of species than others. In this model we saw that patches of ground that were just burned had one set of species, and other patches that hadn't been burned in a while had another set of species. Patchy distributions like this can be formed either through disturbance, such as in this example, through uneven distributions of some resource in the environment, or through interactions between individuals. For instance many animals like to go around in herds, schools, or other groupings, giving patchy distributions of these animals. Among the important creators of patchy distributions in the world today are people, who cut down forests, farm prairies, and so on, leaving only bits and pieces of the native ecosystems. Thus conservation biologists are also very interested in the effects of patchiness and how to deal with them. If you're interested, there are other EcoBeaker labs that explore the effects of disturbance and the effects of patchiness.

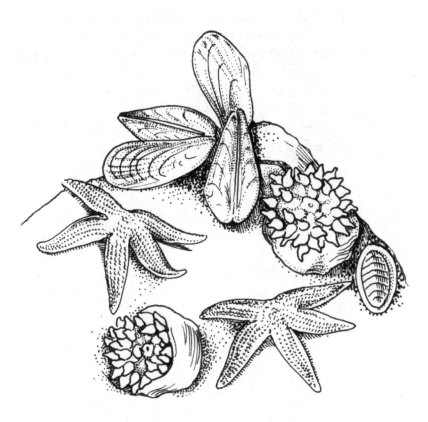

Keystone Predator

Approximate time to complete: 2–4 hours

Background

What determines the structure of an ecological community? One popular method of addressing this question is to make a diagram showing which species eat which other species in that community. A diagram like this is called a *food web*, and from this outline of the community we can start asking more specific questions. Why are there a lot of individuals from one species in a community, while the population of another species is small? Do all species have equal importance to the community, or are some more important than others? Can we predict, just based on the food web, what will happen if one species is removed from the community, or do we need more information? These questions are not only interesting for a naturalist, they are also important in many practical situations, not the least of which is predicting what will happen to ecological communities as more and more species are driven to extinction by human activities.

In this lab we are going to explore these questions by the side of the ocean. If you go to the beach in most places on the west coast of the Americas, or in many other places along the world's oceans, you will find an area called the *rocky intertidal*, lots of rocks that are covered by algae, barnacles, mussels, clams, anemones, and all kinds of other interesting creatures. This habitat is a very good place to study how communities are structured, because it is interesting and complex, yet easily manipulated. The animals move around very slowly or not at all, making counting and observation relatively easy. You can also easily do experiments where you put species in or keep species out of some area of the intertidal, and observe what effects this has on the community. In this lab we'll try to figure out the structure of an intertidal community, using observation and species removals, and in the process we'll discover an important idea in ecology that came from studying the rocky intertidal.

Outline of This Lab

We're going to watch a small area a little way out from the beach in the intertidal area of the coast of Washington State. The rocks in this area contain nine species of plants and animals. There are three species of algae, very simple plant-like creatures that get energy through photosynthesis. *Porphyra* is a leafy green alga often served in Japanese restaurants. *Corralina* is a tougher alga that grows in segments. *Neorhodamela* grows in nifty whorls of tiny finger-like stalks. *Balanus* and *Mitella* are both barnacles, small animals with shells that are shaped like a volcano, with the bottom part cemented to a rock and the top part able to open to let food in and excrement out. *Mytilus* is a mussel, constructed similarly to the mussels and clams you might get in a seafood restaurant. Barnacles and mussels are stationary animals that eat by sticking feather-like extensions into the water and filtering out particles of food, a way of eating that is known as filter feeding.

The last three species are animals that move around and eat some of the other species. *Katherina tunicata* is a chiton, *Nucella* is a snail, and *Pisaster* is a starfish. All of these crawl around slowly through the intertidal, hanging onto the rocks so that they don't get washed away by the waves, and eating anything they consider food that lies in their path.

Among the stationary species, there is a relationship called a *competitive dominance hierarchy*. This means that some of those species can grow on top of others. For instance, a *Corralina* can start growing in an area currently occupied by *Porphyra* and eventually take over all the space that the *Porphyra* was growing in, but a *Porphyra* can't grow on top of a *Corralina*. We say that the *Corralina* is competitively dominant over *Porphyra*.

The goal of this lab is to understand whether the interactions between the different species help to determine which species you see in the intertidal and how abundant each species is. We'll start by trying to figure out which species are competitively dominant to others, then we'll figure out who eats who, building this up into a diagram called a *food web*. Finally, we'll do some experiments in which we take out one species and see what happens to the rest of the community. From these observations and experiments, hopefully we'll be able to see how the intertidal community of species works, and figure out whether some species are more important than others in structuring this community.

Note: While this model is based on the community of species on the outer coast of Washington State, the model is only an approximation. The relationships between the species are reasonably accurate, but the strengths of these interactions, as well as the life history characteristics of the species, are fictional, largely because these numbers have not actually been measured. Despite the inaccuracies, this model behaves in a similar way to the real community when you do the experiments I suggest in this lab.

<Optional> For those of you who are interested in exactly how the model is put together, here is a brief explanation. (If you want more detail on anything described here, refer to the manual for EcoBeaker.) All of the stationary species are modeled using a transition matrix. The way a transition matrix works is as follows. Let's say a certain patch of the intertidal is empty. Over the next year, there is a certain chance that this empty patch will be filled with *Porphyra*. There is also a chance that it will be filled with one of the other algal species, or with a barnacle or mussel. An empty square becoming filled with something is called a transition, and the chance of each transition happening is stored in the transition matrix. Of course, there's also a chance that an empty space will still be empty next year. In EcoBeaker, this is called the No Change transition.

In a similar way, *Porphyra* that is growing in an area this year may be displaced by one of the other algal species next year. So *Porphyra* also has a transition matrix that gives the chance that a *Porphyra* this year will be replaced by each of the other species, stay *Porphyra*, or die (transition to Empty).

I have used the transition matrix to specify the relationships between all the stationary species in the model. Species that are better at settling into empty space have higher transition probabilities in the transition matrix of Empty. If species B can outcompete species A, then the transition matrix of species A will have a nonzero probability for becoming B. The higher the probability, the better B is at outcompeting A. Note that in this model, I made the competitive dominance hierarchy one-way. If B can grow on top of and outcompete A, then A has no chance of outcompeting B. The entry in B's transition matrix for probability of going to A is 0.

The moving predatory species are all modeled with a set of rules called Predator. See the manual for a description of this procedure.

The Lab

1. Run EcoBeaker (double-click on its icon).

2. Open the situation Keystone Predator (use the OPEN command in the File menu).

After the situation loads, you should see several windows laid out on the screen as follows:

A view from above of the area of the intertidal where we'll be working

A graph showing the number of each species currently in the Intertidal Area

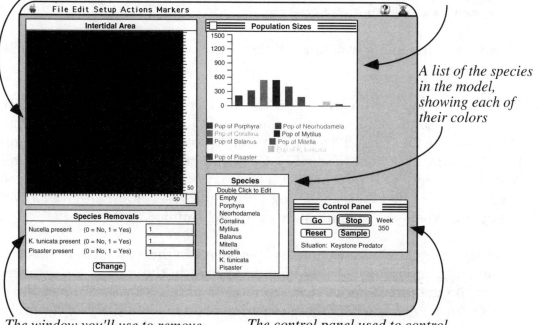

A list of the species in the model, showing each of their colors

The window you'll use to remove species from the model

The control panel used to control running and stopping the model

The big window in the upper left is a view of the Intertidal Area we'll be working in, showing the positions of all the creatures. Each creature has a different color that identifies what species it belongs to, and the key to these colors is given in the window labeled Species. The Species window has a list of all the different species in the model, with each name colored according to the color of that species in the Intertidal Area.

To the right of the Intertidal Area is a graph entitled Population Sizes, showing how many of each species are within the area. Below the Intertidal Area is another window labeled Species Removals, which we'll use later to remove species from the model. Finally, in the lower-right corner is the Control Panel window, which has the controls for running the model.

4. Run the simulation (click on the GO button in the Control Panel).

In the intertidal window you will see *K. tunicata*, *Nucella*, and *Pisasters* start running around, eating. As they eat, they leave behind bare rock (colored black), which will quickly get settled by one of the stationary species. Some of the stationary species will also settle on top of other stationary species and outcompete them. In the graph, you will see that the number of each species in this area goes up and down over time.

5. Spend a few moments watching all this action and getting a feel for what's happening.

6. If you are having trouble distinguishing between some of the species in the model because the colors are too similar, try this. Stop the model (push the STOP button on the Control Panel). Then move your mouse so that it's pointing to the creature you want to identify. Click and hold down the mouse button. A small window will pop up at the bottom of the screen, telling you the species name of that creature, along with some other information that we'll use below. When you're done, start the model running again (GO).

We are going to do some simple experiments in this system to try to determine the importance of the different species in the model. We'll concentrate on the three species that can move around, *Nucella*, *K. tunicata*, and *Pisaster*. Before we start these experiments, we should try to figure out exactly what each of the moving species eats. We should also figure out the competitive dominance hierarchy of all the other species.

To help you determine the competitive dominance hierarchy, I am going to give you part of it in a diagram and then let you finish it:

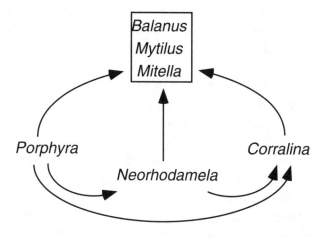

The arrows in this diagram point from poorer to better competitors. For instance, the arrow pointing from *Porphyra* towards *Neorhodamela* shows that if *Porphyra* is growing in some area, *Neorhodamela* can come along and take over that area. Similarly, *Corralina* can take over space from either *Porphyra* or *Neorhodamela*, as indicated by the arrows pointing from the other two algae species to *Corralina*. However, neither of the other two algal species can take over space occupied by *Corralina*, so *Corralina* is the most competitively dominant algae among these three.

As shown by the arrows, any of the filter-feeding species can take over space occupied by any of the algal species. I haven't given you the competitive dominance hierarchy between the filter feeders themselves, though, and this is what you will figure out next.

7. To figure out the competitive dominance hierarchy among the two barnacles and the mussel, we need to do some observations. Find a square containing *Mitella*. Watch that square until something else replaces the *Mitella*. You now know that whatever replaced the *Mitella* is competitively dominant to it. Find several other squares containing *Mitella* and repeat the process until you are fairly sure you have seen everything that is likely to competitively displace *Mitella*. (Note that a *Mitella* getting eaten and a new species moving into the empty square doesn't count towards competitive dominance.) When you are done, make a new diagram like the competitive dominance hierarchy shown above, and add arrows into it showing the new competitive relationships you just discovered.

Now you know the species that are dominant over *Mitella*. In this model (and for the most part in the real intertidal) the competitive dominance hierarchy is one way—*Mitella* will not be able to grow on top of one of the species that you just found was dominant to it. So, in building up the hierarchy, if you see some species displacing *Mitella*, then you can be pretty sure that *Mitella* will never displace that other species.

If you are having trouble seeing your square change colors because the model is running too fast, you can slow the model down as follows. Find the Setup menu, and select OTHER.... A big dialog box will appear. Look near the bottom of this dialog box for an item that says Timesteps/sec. This is the number of weeks the model will run for every second of computer time. Currently, it should be set to 60 (the maximum—though your computer is probably not fast enough to run models that quickly). You can lower this number to slow the model down. For instance, if you set it to 1, then only one week will pass each second you run the model. When you have set it the way you want, click the OK button at the bottom of the dialog box. Later, when you want the model to run quickly again, repeat the procedure and set Timesteps/sec back to 60. You can also make the squares larger by making the window bigger. Click in the lower right-hand corner of the window, hold down the mouse button, and move the mouse to expand the window.

8. Repeat step 7 for the other barnacle species, *Balanus*, and the mussel, *Mytilus*. Draw new arrows in your competitive dominance diagram for new competitive relationships you discover.

Next we want to find out what each of the predator species eat. One trick that ecologists use to find out what creatures have been eating is to look at what's in their guts or in their excrement. If you eat something, it will remain in your stomach for a little while, then whatever parts of it you don't digest are expelled as waste. The partially digested food in the gut is called the gut contents. Normally, when people want to study gut contents they have to kill the animal, cut it open, and look at what's in there. (Can you believe people actually do this as a job?) Within EcoBeaker I have devised a kinder and gentler method for locating and examining gut contents, which we'll use to figure out what each of the predatory species is eating.

9. Stop the model (push the STOP button on the Control Panel). Now find a creature for whom you want to know the gut contents. Move the mouse so that the pointer is on top of this creature and press down and hold the mouse button. A window will pop up on the screen telling you the species name of this

creature and some other information. Among the other information is something labeled Gut Contents, with a number next to it. This number tells you what's in this creature's gut. Note that this will work only for the moving species, not for the stationary ones. The algae don't have guts, and the barnacles and mussels have gut contents that are too mushy to distinguish anything.

> When you get the mushy contents out of an animal's gut, it's not obvious what that mush came from. You have to interpret it a little. In EcoBeaker, the interpretation is easy. Take the number that you got for Gut Contents, look over at the Species window, and count down from the top until you get to your number. The number 1 means that your creature recently ate a *Porphyra*, the number 5 means it recently ate a *Balanus* and so on. 0 means it hasn't eaten anything recently, and its gut is empty.

10. Look at the gut contents of the *K. tunicata*, *Nucella*, and *Pisaster* in our area of the intertidal. As you're gathering the information, write down not only which species are eating which other species, but also the importance of each prey species to each predator. For instance, record what percentage of *Pisasters* have recently eaten a *Neorhodamela*, what percentage recently ate a *Balanus*, and so on.

11. Make a little picture showing which species eat which other species. You can do this in the same manner that you drew the competitive dominance hierarchy. Write down all the species on a piece of paper, and then draw arrows from each species to the other species that eat it. Next to each arrow, write down the relative importance of that prey to that predator (from the percentages you calculated above of predators that recently ate that prey instead of other prey). A simple example of this type of diagram (for illustration only, and not at all related to this model) might look like:

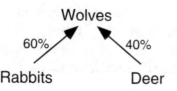

> This picture shows that 60% of the diet of these wolves is rabbits and the other 40% is deer.

12. You may want to run the model for a while longer (GO), then stop it again (STOP) and repeat step 11. This repetition will tell you how constant the percentages you found are, and how much the food that each species eats changes over time.

This diagram you have just drawn, known as a *food web*, is an example of the type of information ecologists gather about the system they are working in. Unless you are a real artist, your food-web drawing probably looks like a real mess, which is what most real-life food webs look like. This makes it a bit hard to draw conclusions from them, although many people have tried.

The next step in this lab is for you to try using your food web to make some guesses about how this community functions. What we're going to do is remove each of the three predatory species in turn and look at what effect its removal has on the whole community. Before we do that, however, you should make some notes on what the community is like now, then try to predict what effect the removals will have.

13. Take some notes on the community as it is now. This is the baseline; you'll compare the results of your experiments to this. Now look at the Population Sizes graph and write down which species are most abundant and which are least abundant. Look at the Intertidal Area and make some notes about what the distribution of species looks like there. You may also want to make notes on how much all of this has been changing as you were running the model (or you could run the model for a little longer and watch how stable it is).

<Optional> For those of you who are interested in being more quantitative in this laboratory, there are several statistics within EcoBeaker that you can use to measure aspects of species richness. For a more quantitative approach, look in the manual to see how to set up a statistics window (it's in the "Graphs" section), and use Simpson's or Shannon's diversity indices, and other statistics that you think might be appropriate.

14. Now, for each of the three species that we'll try removing (*K. tunicata, Nucella,* and *Pisaster*), make a prediction about what's going to happen to the rest of the community after that species is gone. You can predict what will happen to population sizes of each species, what will happen to the distribution of species in the intertidal, or any other characteristic that you think is interesting. If you have quantitative predictions, that's even better. Write these predictions down.

Let's start by taking out *K. tunicata*, the chiton.

15. Find the window at the bottom of the screen labeled Species Removals, and look the item called K. tunicata present. Change the 1 in this box to a 0. Then click on the CHANGE button at the bottom of the Species Removals window.

What you've just done is to remove every *K. tunicata* individual that enters the Intertidal Area as soon as it enters. This is like hiring a research assistant to stand over this little area of beach and watch for *K. tunicata*, and every time she sees one come in, to pick it up and throw it back out. In order to save time, I am not going to explain what we just did in modeling terms, but if you're interested, refer to the transition matrices section in the manual.

16. Now watch for a while longer as the model runs along. What changes do you see? Was your prediction right? Try to be a little quantitative—the population of this species went up around 10%, that one seems to have dropped in half, and so on—and write down these observations. Make sure to watch for a while before continuing to the next step so that the community has time to come to an equilibrium, where the population sizes are more or less stable. A good length of time to wait is perhaps 100 weeks. The Control Panel shows how many weeks have gone by.

17. When you are pretty sure that the community has come to its new equilibrium, and you have taken as many notes on this experiment as you want, it's time to add *K. tunicata* back in. Go to the Species Removals window and change the 0 for K. tunicata present back to a 1. Click on the CHANGE button. Then run the model for a while longer, until the original equilibrium is reestablished.

18. Now let's try removing *Nucella*. The procedure for removing *Nucella* is the same. Change the 1 for Nucella present to a 0, and click on the CHANGE button.

19. Again, let the model run for a while until a new equilibrium is reached. What is this equilibrium like? How is it different from the control situation, and from the situation when *K. tunicata* was removed? How close was your prediction? Try to be a little bit quantitative, and make notes about what you see.

20. When you've figured out what happens upon removing *Nucella*, put it back in. Wait until the original equilibrium is reestablished.

21. Finally, try removing *Pisaster*. This change uses the same technique as the other two. Find Pisaster present in the Species Removals window and change the 1 to a 0. Then click on the CHANGE button.

22. Run the model for a while longer. What's happening now? How is this different from when *Pisaster* was in the model? How is it different from removing one of the other two species?

More Things to Try

If you want to play around more, try removing some of the other species in the model, and seeing what happens then. This is a little more complicated since I haven't set it up for you, but it's not too bad. Find the species you want to remove in the Species window. Double-click on its name (move the mouse to point at its name and double-click). A complicated-looking dialog box will appear. Look in the right side of the dialog box for the set of numbers called the Transition Matrix. Write down all these numbers. Then change them all to 0 except the one labeled Empty, which you should change to 1. Then click on the OK button. That species will now be removed. When you want to put it back in, go back to its transition matrix and type in all the numbers exactly as they were before.

You might also want to try removing two species at once. For instance, you might try removing both *Pisaster* and *Mytilus*.

Perhaps more interesting than doing more removals is to try to figure out when you do and don't get keystone predator behavior (see description below). Think of one change you could make in the transition matrix of one of the species that would make the phenomena of keystone predation disappear, then try it.

Notes and Comments

As you saw in this lab, not all species are equal in an ecosystem. Removing one of the species in the model had a much larger effect on the community than removing either of the other two. This species is called a *keystone species* for this ecosystem, because it, more than any of the other species, determines the structure of the community. This species is the keystone even though it is usually the least, or second-to-least common species when you are running the model. Note that the keystone species in this model is a predator at the top of the food chain. This position is typical of many keystone species, and these *top predators* are often also the most vulnerable to natural or man-made disasters. As you saw in this lab, if the population of these predators is reduced, that change can have significant consequences in the ecosystem, larger than you might guess from the size of the predator population alone.

References

Paine, R. T. 1966. Food web complexity and species diversity. *The American Naturalist* 100: 65–75.

Barnacles and Tides

Approximate time to complete: 1–3 hours

Background

When we tell our children about different species and where they live, we naturally start talking about weather and the physical environment. Camels can go a long time without drinking, so they live in the desert where there isn't much water. Polar bears don't care how cold it gets, so they live in the Arctic where its really cold. A polar bear probably wouldn't be very happy in the desert, and a camel would have a hard time in the Arctic. (Can't you just imagine a herd of camels pulling a sled over the ice?) But is this correlation true for most species? Are most species living in places only because they are adapted to the weather, or might other species that are around also be important?

Several EcoBeaker labs explore the importance of the interactions of species to their distributions. This lab simulates a famous interaction between two species of barnacles on the rocky intertidal coast of Scotland. Barnacles are small animals with shells that are shaped like a little volcano. The shells are cemented to a rock or other hard object, so that an adult barnacle can't move anywhere. Inside the shell is a mouth, out of which comes feathery legs that the barnacle uses to filter the water for food particles. Although the adult barnacle can't move, it makes larvae that can swim around in the water and look for new places to settle down.

Two common species of barnacle live on the Scottish coast, one called *Balanus balanoides* and the other known as *Chthamalus stellatus*. When you walk along the beach, you can see that, above a certain line, the rocks are covered by *Chthamalus* but have very few *Balanus* individuals. However, below this line the frequency is reversed, with *Balanus* covering the lower rocks. In the early 1960s a researcher named Joseph Connell decided to investigate why this was so. He knew that one important determinant of where things live in the intertidal is how often they are exposed to air. Twice a day (on average) the tide goes out and many rocks are left exposed. Species that can't handle exposure to air have trouble surviving on the upper parts of these rocks. Connell wanted to know whether exposure to air was the only condition that governed which species of barnacle lived where, or whether some interaction between the two species of barnacles also had something to do with it. In this lab, we'll repeat some of Connell's experiments.

Outline of This Lab

In this lab, we will be looking at the side of a large rock somewhere along the coast of Scotland. The rock has two species of barnacles on it, *Chthamalus* and *Balanus*, as described above. New individuals of both barnacle species are continually trying to settle down on the rock surface, and in this model, once they settle, they grow up right away. None of the barnacles in the model die of old age, but they can die of several other causes. One way a barnacle can die is by being exposed to air for too long. In the model, every day the tide goes down, exposing barnacles in the upper part of the rock to air, and this exposure may be able to kill barnacles. It's also possible that barnacles could die from being submerged underwater for too long, and never getting exposed to air. In addition, barnacles may die from competition with other barnacles, as larvae may be able to settle next to or on top of an adult barnacle that's already there, and then grow over the top of it. Some combination of these effects is responsible for the distribution of the two species of barnacles on our rock wall.

Your job is to come up with a set of hypotheses about why *Chthamalus* lives high on the rock and *Balanus* lives lower down. Then you should design and carry out some experiments to distinguish between these hypotheses. I will tell you about some tools in the program that you can use to do your experiments, and also make some suggestions based on Connell's studies.

The Lab

1. Run EcoBeaker (double-click on its icon).

2. Open the situation Barnacles (use the OPEN command in the File menu).

After the situation loads, you should see several windows laid out on the screen as follows:

A view of the rock on which we'll   *A list of the species in the model,*
be conducting experiments   *showing each of their colors*

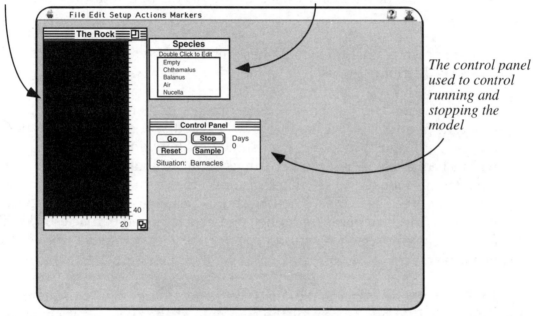

*The control panel
used to control
running and
stopping the
model*

On the left is a window labeled The Rock, giving a view of the rock surface where we'll do our experiments. To start the model, we'll wipe off a piece of rock so there are no creatures on it at all, only seawater (colored blue). The piece of rock we picked is perfectly vertical in the water, and its top is just at the level of the high tide, so at high tide the rock is completely covered by water. You can see that right now it's at high tide and the water goes up to the top of the rock. To the right of The Rock are two windows, one labeled Species, which shows all the species in the model and their colors, the other labeled Control Panel, which is used to control starting and stopping the model.

3. Run the simulation (push the GO button in the Control Panel).

You will see tan *Chthamalus* barnacles and black *Balanus* barnacles starting to settle on the rock. Each tan or black square that appears is a barnacle that has settled down. Every day the tide comes in and out, and this is represented by air (colored white) coming down from the top of the rock and displacing the blue water. The height of the tide has a 28-day cycle, so that for each of 14 days the low tide will get successively lower and lower (the white air will come farther down the screen before receding), and then for the next 14 days the tide will go back up. This tidal cycle comes about because tides are caused by the moon—as the moon waxes and wanes, the tides get larger and smaller.

4. Watch the simulation until you see a very clear distribution of the two barnacle species. Write down a short description of this distribution and how it relates to the tide.

5. Now write down at least three hypotheses about why you see the two species of barnacles distribute themselves as they do.

You now want to do a series of experiments to distinguish among your different hypotheses. To help you, let me tell you some of the types of things that Connell did, and then show you how you can do those from within EcoBeaker. Among other things, Connell looked for dying barnacles, and also looked to see whether individual barnacles were ever overgrown by other barnacles. Then he did experiments in which he moved barnacles from one place to another, and also removed barnacles from different areas of rocks. You can do each of those experiments from within EcoBeaker.

To see whether a barnacle dies or is overgrown, simply pick a square with a barnacle in it and watch for the square to change color. If it changes to blue, that means that barnacle died for some reason. If it changes to the color of the other species of barnacle, that means that the original barnacle was overgrown by an individual from the second species of barnacle.

You can't quite transplant barnacles within EcoBeaker, but you can place new barnacles in any position you want. To do this, first stop the model (push the STOP button in the Control Panel). Next go to the Action menu and select PAINT. This puts you into Painting Mode (the Control Panel will tell you what mode you're in). Now pick the species that you want to transplant by clicking once on its name in the Species window. That species name should now be highlighted in the Species window. Finally, move the mouse so it points at the square on the rock where you want to add this species, and click once. That square will change color, indicating that it now contains an individual of the species you wanted to add. When you are done adding in individuals, go to the Control Panel and click on STOP. This will return you back to EcoBeaker's regular mode, where you can start the model running again.

To remove species from a certain area, you use the same procedure, except now select Water as the species which you want to paint on. Then click on the individual you want to remove, and that square will change to water, with no barnacle there. You can hold down the mouse button and drag the mouse to remove barnacles from a whole area of the rock. Again, push the STOP button in the Control Panel when you're done to return you to the normal mode in EcoBeaker. To keep a species removed for a period of time, you may need to repeat this process every so often.

Finally, here are two more tools that you may find convenient, one that counts barnacles for you and another that slows down how fast the model runs. If you push the SAMPLE button on the Control Panel, a line will be drawn across The Rock at some height, and EcoBeaker will ask whether you want to sample the species along that line. If you say yes, then EcoBeaker will count the number of each species under that line and report it to you. If you say no, then EcoBeaker will move the line to a different location and ask again. You might use this as a way of measuring species abundance at different heights.

During some of your experiments you may want to slow down how fast the model is running. To do this, go to the Setup menu and select OTHER. Ignore everything here except the box down near the bottom labeled Timesteps/sec. This field shows how many days go by for every second you run the model. The smaller the number you put in here, the slower the model will run.

6. Design a series of experiments to test your hypotheses, and then carry out these experiments. You might want to do each experiment a few times just to make sure that you didn't get some result by chance. Feel free to change your plan of experiments based on the results you get.

7. Were any of your hypotheses right? Was some combination of them correct? Are you not yet sure? Write down your explanation for what is determining the distribution of barnacles in this system. If you're not yet sure, write down as much as you know, then suggest further experiments that you could use to discover the rest of the story.

More Things to Try

You may have noticed that there is a third species in the model called *Thais*, which we haven't looked at yet. *Thais lapillus* is a species of snail that crawls around below the level of the tide eating barnacles. It does not exist in all locations along the Scottish coast, but where it does exist it can have a large effect on the distribution of barnacles. To see this, try adding in the snails. You can do this the same way you added in barnacles, by painting in a snail or two.

Watch what happens for a little while. Then you can again come up with hypotheses as to why the snail has the effect it does, and test these hypotheses with similar experiments as you used above. For a further exploration of the type of effect that the snail has here, see the Keystone Predator lab.

Notes and Comments

The intertidal area of the ocean has been a really useful place to find out how species interact with each other and with their physical environment, and how these interactions determine which species are found in which places. The intertidal is an interesting place in and of itself, and can also be important economically. For instance, that's where you find a lot of seafood. However, the intertidal is only a very small strip along the world's coasts, so perhaps the general concepts that emerge from intertidal studies are more important than the specific results. Connell's work with barnacles provided some of the first really good evidence that competition between species can be important in a natural situation. It also showed very clearly that there are trade-offs in the way species are constructed. One barnacle may be able to outcompete another, but can't survive in as many environments. This type of trade-off is quite common. You never find a single species that is good at everything.

Another thing we learn from this lab is that looking is not enough. You must also do experiments. Without the experiments that we did here, we couldn't know whether the two species of barnacles lived where they did simply because one liked air and the other liked water, or whether they actually competed with each other. Only through the experiments were we able to figure out which hypothesis is right. This is true in general—it's very hard to do ecology without experiments, and if you don't experiment, then many times you get the wrong answer.

If you're interested, EcoBeaker includes several other labs looking at the interaction between species, and one of these also looks at interactions in the intertidal area. For instance, you can take a look at the Keystone Predator lab for more work in the intertidal, and at the Competitive Exclusion lab to look at species interactions in a general way.

References

Connell, J. H. 1961. The influence of interspecific competition and other factors on the distribution of the barnacle *Chthamalus stellatus*. *Ecology* 42: 710–723.

Island Biogeography

Approximate time to complete: 2–5 hours

Background

During the heyday of European exploration of the world in the sixteenth through nineteenth centuries, countries such as Spain, France, England, and the Netherlands were sending out sailing ships for months or years at a time. You can imagine that spending months at sea on one of these ships could get pretty boring, even if you were a veteran sailor with lots of other sailors around to fraternize with. How much more boring it must have been to be one of the captains, who (in some countries at least) were from a rather snobby class and wouldn't deign to socialize with their crew. To combat this loneliness, captains of these ships would take along a naturalist or two, someone from the same social class as themselves, who would provide company and conversation while on the high seas, and who would also be marginally useful once they got to land. If your exploration didn't find gold, at least your naturalist might find some interesting new type of animal to bring back. In this way, many scientists in the eighteenth century got to travel around the world; the most famous of these was Charles Darwin on the *Beagle*.

These scientists landed on many islands all over the place, as well as along the shores of the continents, and wherever they went they found and catalogued new species. One of the main questions they started asking was, Why do you find certain species in certain places and not others ? Place in this case refers to a large area, like a country or an island. This question has expanded into a field called *biogeography*, a part of ecology that tries to figure out what types of species live in different areas of the world, and why each species lives where it does.

Among the interesting observations biogeographers have made is that there are an incredibly large number of different species on continents, a really large number of species on big islands, a smaller number of species on medium-sized islands, and, in general, not all that many different species on really small islands. They also have noticed that the farther away an island is from a continent or another big island, the fewer species will be found on it. People have come up with several theories to explain these observations. It might be hard for some species that can't fly or swim to cross the ocean and get to an island. It might also be harder for many species to coexist on a small island, perhaps because there are not enough different kinds of habitat on an island to support many species. All of these theories have some merit, and are no doubt partially responsible for there being fewer species on islands than one would expect.

In the late 1960s, Robert MacArthur and E. O. Wilson combined some of these ideas and came up with another hypothesis, which has been quite influential. They thought of things this way: If you start with an island without any life on it, in a short time individuals of some species will reach it. Initially, every individual that reaches the island will be a new species, and so the rate of species colonization will be very high. However, as time goes on, new individuals coming to the island will probably be members of species that are already established on the island, so the species colonization rate will go down, until finally every species around will be represented on the island and the species colonization rate will be 0. At the same time, if we look at the rate of species extinction on the island, this rate will start out at 0, since on an empty island there are no species to go extinct. As the number of species on the island starts rising, the number of species going extinct on the island will also rise. At some point, these two curves will cross, and that is the *equilibrium number of species* for that island—that's the number of species you'd expect to find on that island. You can see this idea represented graphically as follows:

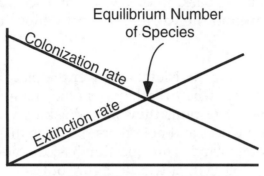

The two lines show how the colonization and extinction rates change as the number of species on the island increases. The place where the two lines cross shows the number of species you would expect to find on the island.

Both the species colonization and extinction rates will be affected by the size of the island and its distance from the mainland. The farther an island is from the mainland, the harder it is for species to get there, so the lower the species colonization rate will be. On the above graph, the colonization line will be moved downward. Smaller islands may be harder to find, so a smaller island might also have a lower colonization rate. A larger island will be able to support larger populations of each species that reaches it, so the chance that a species will go extinct through bad luck will be less on a larger island and the extinction rate should be lower. On the above graph, the extinction line will be moved downward. Similarly, a larger island may have more niches than a smaller island, allowing more species to coexist, thus leading to a smaller extinction rate.

In this lab, we'll test this idea to see if it works for a simple population of birds. We'll have a number of bird species on a mainland. Then we'll change the position and size of an island and look at how many species end up on it.

Outline of This Lab

In this lab, I've set up a little world showing the edge of a continent, an ocean, and an island in the ocean. Eight species of birds live on the continent, with the scientific names Purple Bird, Red Bird, Blue Bird, etc. The birds live on packets of Bird Seed that drop out of the sky onto the mainland and the island every day. You can think of these as generic birds that you might find on any coast, and the Bird Seed as a proxy for whatever food those birds eat.

These are slightly odd birds. Each bird starts out life with a certain amount of energy—you can think of this as how much fat they have. They live off of this energy, and everyday they use up a little of it. The birds fly around at a certain speed looking for Bird Seed, and each time they eat a packet of Bird Seed they gain more energy. If a bird can't find enough Bird Seed and runs out of energy, it dies. However, if it can find enough Bird Seed to eat it can build up energy; when it accumulates a certain level of energy, it will have a baby bird. The bird gives half its energy to this baby and keeps the other half for itself. To start with, we'll make each of the species of birds exactly the same as the others. This means they all fly at the same speed, have baby birds at the same energy level, and so on.

Being the exceptionally strong and smart people that we are, we can pick up the island in this world, move it around to wherever we want it, stretch it out into a large island or smoosh it down into a small one. In this way, we can have the island vary in size and distance from the mainland. For each island we can then watch how many species of birds end up on it, and look for new immigrations and extinctions. These experiments should let us explore whether the number of species on an island can be related to immigration and extinction, and whether we reach some sort of equilibrium number of species this way. We can also play with the different species of birds, making some better able to cross the ocean than others, and see if that changes the results.

Note: This lab can get a little bit more complicated than some of the other labs in EcoBeaker, so you may want to try one of the other labs first.

The Lab

1. Run EcoBeaker (double-click on its icon).

2. Load in the situation file Island Biogeography (use the OPEN command in the File menu).

After the situation loads, you should see several windows laid out on the screen as follows:

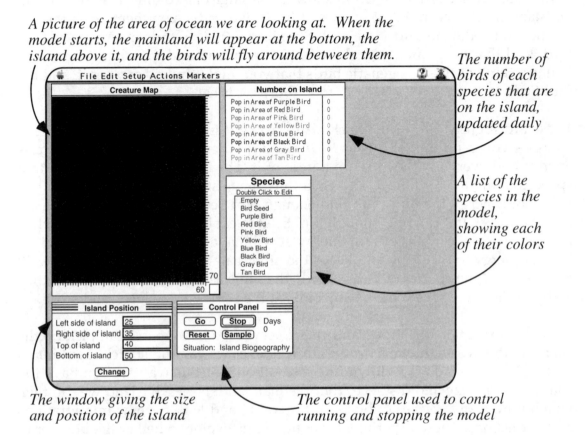

A picture of the area of ocean we are looking at. When the model starts, the mainland will appear at the bottom, the island above it, and the birds will fly around between them.

The number of birds of each species that are on the island, updated daily

A list of the species in the model, showing each of their colors

The window giving the size and position of the island

The control panel used to control running and stopping the model

There are a lot of windows on the screen, so let me just take a minute to tell you what each of them shows. The biggest window in the upper left labeled Creature Map, shows the positions of the mainland, the island, all the birds, and the green packets of Bird Seed in this world. The blue is ocean. When you start the model running you'll see dark brown land masses appear, along with green packets of Bird Seed, and other colored squares, which are the birds themselves. The scales for this map are shown on the right and along the bottom. Notice that the upper-left corner is square (1,1), and the lower-right corner is square (60, 70).

Below the Creature Map are the Control Panel and a window called Island Position. The Control Panel lets you start and stop the model. Also on the Control Panel is how many days have gone by since you started the model running. The Island Position window is where you can set the position and size of the island, which I'll describe below.

Next to the Creature Map is a window called Number on Island. This window shows the number of birds of each species that are on the island each day.

Finally, the Species window lists all the species in the model and shows you their colors. Here you can see that the little red squares on the Creature Map are birds from the species Red Birds, the little yellow squares are from Yellow Birds, and so on.

3. Run the model (press the GO button in the CONTROL PANEL).

Notice that the mainland appears at the bottom of the Map, and an island appears above it. Green packets of bird seed start falling down on both the mainland and the island; birds appear on the mainland and start flying around, eating bird seed, reproducing, and dying.

4. When you've seen enough, stop the model (push the STOP button on the Control Panel).

The first thing to do is see whether in this model we find the same thing that the biogeographers found. Do we get fewer species of bird on smaller islands, and do we get fewer species of bird on islands that are farther away from the mainland? To check, we can make islands that are different sizes, and different distances from the mainland, and see how many species end up on each island.

5. Start the model running again (GO) and figure out how many species of birds are on this island.

As you'll discover, the number of species on the island is not constant over time, so you'll have to watch for a while and then take some sort of average. You can use the Number on Island graph to help you see exactly how many species are on the island if it's hard to see on the Creature Map. Don't just make a guess at the average—decide on some way to count the number of species on the island at different times, then average those counts together.

6. When you have figured out an average number of species on this island, write that number down along with the size and position of the island (shown in the Island Position window), and then stop the model (STOP).

7. Now change either the size of the island or its distance from the mainland. To change the position of the island, find the Island Position window, and type in the left, right, top, and bottom coordinates of your new island position. Remember, the scales along the sides of the Creature Map shows you the coordinates. As you can see, the top of the mainland is at square 60, and the island currently extends left to right from square 25 to square 35, and top to bottom from squares 40 to 50.

8. When you've typed in new coordinates, reset the model (click on the RESET button in the Control Panel). Then run the model again (GO).

9. Wait for the birds to appear on the mainland, and then, using the same algorithm as above in step 5, calculate the average number of bird species on the island.

10. Repeat steps 6 through 9 for a number of different island sizes and distances from the mainland.

    Does the model work the way you expected it to? Does island size have an effect? Does distance from the mainland have an effect? You might want to make graphs of your numbers to see any effects more clearly.

    Now you're ready to look at why we get different numbers of species on the different islands. I am not going to tell you how to do this. Instead, let me tell you about some approaches you might want to take, and then I'll guide you through the set of tools you can use within EcoBeaker.

    One thing you may want to do is look at the rates of immigration and extinction of different islands. The immigration rate shows how many new species of birds come to an island per unit of time, and the extinction rate shows how many species of birds that were on the island go extinct per unit of time. For instance, if an island had some Red Birds, some Black Birds, and some Blue Birds on it at Day 100, and by Day 110 it had Red Birds, Blue Birds, and Orange Birds, but no Black Birds, then the immigration and extinction rates over those 10 days were both 1 species. You might also want to combine these rates into some sort of species turnover rate.

    To calculate these rates, you might just write down on a sheet of paper every time a new species comes to the island and every time a species goes extinct on the island. Alternatively, you might do more controlled experiments such as placing birds from a certain number of species on an island, then running the model a fixed number of days and seeing how many are left. Here's a guide to doing an experiment like that.

11. Place the island where you want it using the Island Position window, and then RESET the model. Run the model for a short time, so that birds start appearing on the mainland. Then STOP the model.

12. Now go to the Action menu and select the PAINT command. First, we'll get rid of any creatures that might have already reached the island. Find the Species window, and click once on the name EMPTY. Empty should now be highlighted. Then move the mouse over to the Creature Map and click once on every bird you see that is on or next to the island. Each bird you click on should disappear. You are painting over these birds with empty squares.

Now we'll add in the birds we want on the island. Let's say we want three species of birds on the island, and we just randomly pick these three species to be Black, Red, and Purple Birds.

13. Stay in Painting mode, and move the pointer over to the Species window. Click once on one of the bird species that you want to add to the island. That species should now be highlighted. Now move the mouse back over to the Creature Map, position it around where the island is, and click once. You will see one bird from the species you selected placed on the island. If you want to place two birds of this species on the island, move the mouse to a different place on the island and click the button again.

14. When you are done adding birds from this species, go back to the Species window and click once on the next species you want to add. Then move the pointer back to the Creature Map window and click in the places where you want birds of the second species. Repeat this for each species of bird you want to add.

15. When you are done adding birds, click on the STOP button in the Control Panel. Now you are back to the regular control panel, and you can run the model as before.

At this point, you might want to run the model for a certain number of days, and then see how many of the original species went extinct and how many new species immigrated. This experiment would give you rates of extinction and immigration for that island, with that number of initial species. Since there are a lot of chance events involved here, you should probably repeat this several times and average the numbers. Also remember that some of the theories predict that colonization and extinction rates will depend on the number of species on the island, so you should probably try adding different numbers of species to each island and seeing what rates you get. You may also want to try making graphs of the extinction and colonization rates versus initial number of species on the island.

You may also want to try making some species of birds better than others at getting to islands. I will give a brief description of how to do that here, but you should refer to the manual if you haven't done this type of thing before in EcoBeaker. The most obvious three characteristics that you could change to make birds from a certain species better colonizers are how fast they fly, how much they change direction while flying, and how long they can fly before they run out of energy.

As an example, let's say you wanted to make Black Birds faster and able to fly more days before running out of energy. Here are some instructions for changing those things.

16. Find Black Birds in the Species window, and double-click on it. This should bring up a complicated-looking dialog box. There is an explanation of the entire box in the manual. For now, just look in the lower-left corner of the dialog box for the button labeled Action Params and click on it. This will bring up a second dialog box, where you can set the parameters for Black Birds.

Here's a quick description of the different parameters you see here. A more detailed description is given in the manual; you should look there if you're uncomfortable with this quick version. The bird can look for food at Look Distance squares per day. If it sees some food, it moves towards it at Speed squares per day. If it doesn't see any food, it wanders around randomly for Num Random Steps days. If it still hasn't found any food, it then picks a direction and keeps moving in that direction until it encounters some food. This direction and speed can be modified by up to Move Randomness squares every day. A value of zero for Move Randomness means that once a bird picks a direction, it keeps going in exactly the same direction with the same speed, and numbers greater than 0 mean the speed and direction will change over time. Each time it eats something, it gets Prey Value units of energy; it uses up Cost of Living energy units every day just surviving and flying around. When it reaches Repro Energy units of energy, it reproduces. The last parameter, Prey Species, gives a list of which species this species of bird can eat. You can change that list if you want to make some birds eat others.

17. To change the speed of Black Birds, find the Speed parameter and change it to 4. To make Black Birds more hardy, let's reduce how much energy they spend maintaining themselves. Set Cost of Living to 0.5. Then click on OK in both dialog boxes. Black Birds are now faster and hardier than the others.

> One last thing. If you are comfortable manipulating files on a Macintosh and using a spreadsheet, there is a trick that will make counting the number of species on the island easier. You can have the Number on Island graph save its data every time it updates to a file, which you can later load into a word-processor or spreadsheet. To do this, double-click in the middle of the graph. A complicated-looking dialog box will appear. Find the button at the bottom labeled Advanced Stuff and click on it. Another dialog box will appear, where you can tell the graph to save its data, and give a file name where that data should go. Check the SAVE DATA FROM GRAPH check-box, and then click on the SAVE FILE button and give a file name. Now all data that is shown in the graph will also be saved to the file you specified, and you can open that file in a spreadsheet or graphing program. You may want to change the file name each time you do a new run of the model so you can keep track of your data. You may also want to make the graph update less often than every day to reduce that amount of data you must deal with. Do this in the graph setup box by clicking on the UPDATE RATE button. All this is explained in further detail in the manual.

18. Go for it.

Notes and Comments

In this lab we looked at islands as you normally think of them—chunks of land in the middle of water. However, many of the same concepts would apply if we reversed viewpoints and thought of lakes as "islands" of water in a "sea" of land. In fact, you can think of many things as islands. Mountaintops are islands in a sea of valleys. Cities are islands in a sea of farms and forests (or maybe it's starting to be the other way around). You can even think of islands on a much smaller scale. In a field of grass with a few blueberry bushes, the bushes might essentially be islands in a sea of grass if you were some little bug that could only live on the bushes and not in the grass. So the ideas that we've seen in this lab can be applied quite broadly.

One important place where the ideas of island biogeography are being applied currently is in conservation biology. Humans have chopped up vast forests, prairies, and other native habitats into little chunks, with roads, cities, and farms in-between the chunks. So for many species, their habitats are now divided into little "islands," and they have to cross unfriendly territory to get from one place they can live to the next. When we think about how best to preserve these species, we use a lot of the concepts that we explored above. For instance, if you have a choice, how far away should you put the islands of habitat from each other? Should you put little stepping-stone chunks of habitat between the bigger chunks? Would it be better to save one big island of habitat, or many smaller islands? Some conservation biologists use models very similar to the one we used here to try and answer these questions, as well as doing experiments along the same lines. Other EcoBeaker labs, including the Corridors, Stepping Stones, and Butterflies lab, explore these concepts further.

References

MacArthur, R. H. and E. O. Wilson. 1967. *The Theory of Island Biogeography*. Princeton University Press, Princeton, N.J.

Corridors, Stepping Stones, and Butterflies

Approximate time to complete: 4–6 hours

Background

Large parts of the western side of the state of Oregon used to be covered by prairies. These prairies had many different plants growing in them, and lots of insects and other animals lived on these plants. Among all of these species was a butterfly called the Fender's blue butterfly. This is a small butterfly, a couple centimeters across the wings, and the males have a tint of blue that gives the butterfly its name. The Fender's blue butterfly needs a certain kind of flowering plant called a Kincaid's lupine to live, and this lupine only grows in relatively pristine prairies.

Over the last one hundred years, most of the prairie in Oregon has been replaced by farms, cities, and roads, and what prairie is left occurs in isolated patches that can be several kilometers away from each other. Because there is so little Kincaid's lupine left, the Fender's blue butterfly is in danger of going extinct. Less than four thousand Fender's blue butterflies are still alive, spread among a dozen or so small patches of prairie that have survived relatively unscathed by the progress of man.

People in Oregon are interested in saving this species of butterfly, along with other endangered prairie species. Because the butterfly is rare, the government and private agencies are willing to spend some money to try to save it. There are old farmlands and other abandoned areas in this part of Oregon that potentially could be converted back into habitat where the butterflies could live. However, converting farmland back into prairie is expensive and takes a lot of time, so we'd like to know beforehand which pieces of land would be most useful to restore.

Conservation biologists generally consider three strategies for restoring or acquiring habitat in situation such as that faced by the Fender's blue butterfly. One is to add habitat onto already existing patches, thereby making bigger patches. Another is to connect patches together by strips of habitat, called *corridors*, with the idea that the species you are trying to save can use the corridor to get from one patch to another. A third way to add habitat is in little patches in between the big ones, called *stepping stones*. Here the hope is that a species can go from one stepping stone to another until it makes it to another large patch. Each of these strategies will be good for some species, but not so good for others.

In order to choose a strategy for the butterflies, it would be nice if we could do experiments in which we could try out different configurations of prairie, put some butterflies in each configuration, and see which arrangement of patches worked best. We can't do this, though, first, because it would require a very large experiment, and second, because the butterflies are about to go extinct, so we don't want to jeopardize even a few of them for our experiments. Because we can't conduct real experiments, the next best thing is to turn to models of the system. Models can't tell us exactly what will happen, but if we use them correctly, they can help us decide which strategy is most likely to work.

In this lab I've constructed a model of Fender's blue butterflies in an area around Eugene, Oregon. I've given you a map of several of the patches of prairie that still exist and that support these butterflies. You will take this map and try adding in more habitat in various configurations, run the model with each configuration, and see how well the butterflies do. Using this data, you'll recommend a restoration plan for the Fender's blue butterfly. You'll write this plan up as described below, and present it as your recommendation to the City of Eugene.

Outline of This Lab

We'll start this lab by taking a time machine back 100 years into the past, when prairie covered a large part of western Oregon and the Fender's blue butterfly was happily going about its business. We'll watch the butterfly do its thing here for a few years. As you'll see, the butterflies reproduce once a year, having about six offspring each that survive to adulthood. They lay many more eggs, but most die before reaching adulthood. Butterflies reproduce only if they are in prairie. The butterflies like to spread themselves out a little, so if a butterfly is too near another butterfly, you'll see it move away. Each butterfly has a chance of death of about 0.4% per week.

After seeing what life was like 100 years ago, we'll take our time machine back to the present. Now, instead of a giant field of prairie, there will be only a few little patches of prairie scattered around a farming area. Each patch will have a few butterflies in it to start out, and you'll see that as the butterflies move around, some of them will leave the prairie patches. Once they leave a patch, they keep flying in a random way until they reach another patch. From previous work, you know that the death rate of the butterflies increases tenfold when they are on farmland instead of on prairie. The reason might be because they can't find enough to eat, they are easier for predators to catch, or they just get so tired of seeing cornstalks that they die of boredom. You also know that just the act of flying around stresses out the butterflies and makes them die faster (kind of like flying some commuter airlines).

We have enough money to buy up a certain amount of land and restore it to native prairie. You'll use the model to test different ways of doing this. Your goal is to keep as high a population of butterflies around as you can for as long as possible.

The Lab

1. Run EcoBeaker (double-click on its icon).

2. Load in the situation file "Butterflies" (use the OPEN command in the File menu).

    After the situation loads, you should see several windows laid out on the screen as follows:

A map of part of western Oregon where the butterflies live

A list of the habitats in the model, showing each of their colors

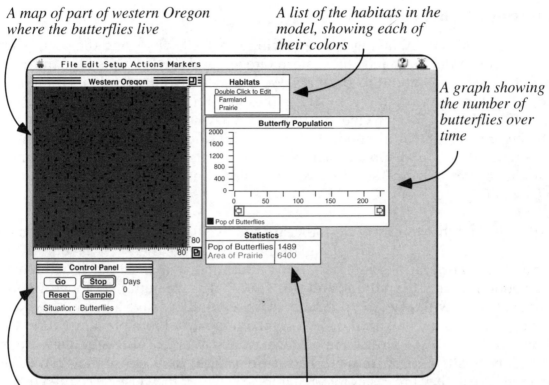

A graph showing the number of butterflies over time

The control panel used to control running and stopping the model

The number of hectares of 100m by 100m squares of prairie and farmland in the map of Western Oregon

The big window in the upper left of the screen is a map of a piece of western Oregon. The map is 8 km by 8 km, and each square in the map represents a piece of land 100 meters on each side. You are looking at the map as it was 100 years ago, completely covered by green prairie. On this prairie you can see quite a few blue butterflies, which will start moving around as soon as you run the model below. To the right of the Western Oregon map are three windows that give you more information about what's happening in the model. The window on top labeled Habitats shows the colors of farmland and prairie on the map. The middle window, Butterfly Population, gives a graph of the number of butterflies on the map every week. The bottom window, Statistics, tells you exactly how many butterflies and how many squares of prairie are on the map at this time. Finally, underneath the Western Oregon map is the Control Panel, which we'll use to run the model. The Control Panel also tells you how many weeks have gone by since you started running the model.

3. Run the model (press the GO button in the Control Panel).

You'll notice the butterflies start to fly around, and gradually this year's butterflies will die off. When the model gets to 52 weeks (as shown on the Control Panel), all the butterflies will reproduce, and they'll reproduce every 52 weeks thereafter. The Butterfly Population graph shows this cycle more clearly.

4. Watch the model run for several years (of modeling time, not your time). Look at both the map and the graph to get a feel for how what the butterfly population is doing over time. Write down your impressions. You may also want to write down some numbers.

5. Stop the model from running (push the STOP button on the Control Panel).

OK, now we're ready to take our time machine back to the present. The butterflies will be exactly the same as they were 100 years ago, but there will be much less prairie.

6. Find the menu labeled Markers, and select the item TODAY from this menu.

You'll see all kinds of windows opening and closing and rearranging on the screen. Wait until it all settles down, then look at the map. This is a map of the same area of Oregon as before, but now you can see that most of the land is tan farmland, with only a few green patches of prairie left. There are still some blue butterflies in these patches of prairie. These butterflies still behave exactly like the butterflies of 100 years ago.

If you look in the Statistics you will see that currently 800 squares of the map are prairie. You have squeezed out enough money from the city of Eugene and various conservation organizations to buy 200 more squares of land and restore them to prairie. In the real world, restoring prairie means slogging around in the pouring rain or the baking sun, painstakingly putting in and digging up plants (or more commonly, finding a few unwitting students to do it for you). In EcoBeaker, it's a little easier. Let me take you through the procedure you use in EcoBeaker to add prairie to the model.

7. Make sure the model is not running. If it's running, stop it (STOP).

8. Find the Actions menu and select PAINT. The Control Panel will now indicate that you are in Painting Mode.

9. Find the window called Habitats to the right of the map of Oregon. Click once on PRAIRIE. Prairie should become highlighted.

10. Now move the mouse to an area of the map of Oregon that you want to restore to prairie. Click on the mouse button. You will see a new square of prairie appear beneath the pointer. You will also see that the Statistics window now says that you have 801 squares of prairie. You can add prairie in more than one square at a time by clicking and holding down the mouse button, then dragging the mouse to outline a rectangular area, and finally releasing the mouse button. The whole outlined rectangle will become prairie.

11. If you make a mistake while putting in prairie, you can correct your mistake by selecting FARMLAND from the Habitat window, and then painting over your mistake with the farmland. You should be careful when you do this, however, not to paint over any of the original patches.

12. When you are done adding in prairie, click on the STOP button in the Control Panel. The Control Panel will then revert back to how it was originally, and you can continue running the model as before.

> Remember, as you are adding in prairie, to watch the Statistics window and stop when you have added 200 more squares (so there will be 1000 squares total prairie). If you add more than this, your restoration plan will be rejected as costing too much.

> If you want to go back to the original configuration of prairie (what's there now), you can either click on the RESET button in the Control Panel, or select the TODAY item from the Markers menu, as in step 6.

> There are a couple more tools in EcoBeaker that you may find convenient. One thing you may want to do is save a configuration of habitat that you want to come back to later. You can save everything as it appears on the screen by going to the Markers menu and selecting the ADD MARKER item. When it asks you for a name, assign some name for this arrangement of habitat that you can remember later on. Now, whenever you want to go back to that configuration, go to the Markers menu, look for the name that you gave that configuration, and select it. It's particularly convenient to do this just after you have added in habitat, but before you start running the model, so you can get back to that restoration plan with all the initial butterflies. If you pile up a whole bunch of these markers, you can get rid of some of them using the Delete Marker item in the Markers menu.

You may want to print out some of the maps you come up with, or the graph of population over time. To print out the map, click once anywhere in the map window, then go to the File menu and select PRINT. Assuming you have a printer hooked up, this will print the map. Similarly, to print the graph, click once in the graph window, and then select PRINT.

Finally, if you need to stop in the middle of your modeling adventure to do something else, you can save what's currently on the screen and all your markers by going to the File menu and selecting the SAVE SITUATION item. Give a file name for the file to save the model in (you should change it from the default), and answer yes to both of the dialog boxes that appear afterwards. You can then load this file back into EcoBeaker at a later time exactly as you loaded the original model in step 2.

Here are a few hints to get you started. Reports are more convincing when there are numbers in them rather than just impressions and guesses. You shouldn't use numbers from a model like this as absolute amounts, but rather as a way to compare between different runs of the model. Also note that what happens in the model is random, so what you see in one run or one year might not happen a second time, even with the same starting conditions.

13. When you are done, write up a report on your results, including your recommendation for the optimal way to restore prairie. The report should be addressed to the City of Eugene, Oregon, and be two pages in length plus any graphs you feel inclined to include. You should also submit a print-out of your recommended prairie restoration plan, with (a) verification that you did not add more than 200 units of prairie and (b) an estimate of the butterfly population with your "best" restoration plan. For this population estimate, run the model for at least 50 years, and then print out a graph of the population over the last few years.

Notes and Comments

This laboratory is modeled after real work that is being performed by Cheryl Schultz, a graduate student in the Department of Zoology, University of Washington. She is gathering data about the Fender's blue butterfly and its habitat, which she hopes to put into a model very much like this one. Using that model, and playing with it in the same sorts of ways as we have done here, she plans to recommend a restoration plan to the City of Eugene, the Nature Conservancy, and the Bureau of Land Management, all of which are willing to put money into helping the Fender's blue butterfly survive. So this whole exercise is actually something people use in dealing with real conservation problems.

This laboratory is in some ways an optimistic conservation biology lab. A more pessimistic lab would have looked at the same issues we explored here, but would have focused on getting rid of habitat instead of restoring habitat. For instance, in the case of another endangered species in the northwest of the United States—the spotted owl—the question is not how to restore its habitat (which is old growth forest and takes forever to grow), but how much more of its habitat we can cut down without the owl going extinct. In this case, the object would be to find a pattern of cutting that minimizes the risk of the owl going extinct. All the same ideas still apply. You might still try to leave corridors of trees between larger forested areas, or leave little patches of forest in between the big patches, or try to make the patches of forest that you leave behind as big as possible. This story is repeated for many other species. The answer for each species will depend on the behavior of that species, and how it interacts with its environment.

Things become even more complicated when you start thinking about more than one species. The Fender's blue butterfly is not the only species in danger of extinction because of disappearing prairies, and the other species that live in these prairies may respond differently to the restoration effort than the butterfly will. Having two of these species interacting with each other directly, such as a predator and its prey, may completely change the way you need to lay out habitat.

Finally, I should tell you that, while the results you got from this model do apply to some species, I have not modeled Fender's blue butterflies closely enough so that the results will apply directly to them. Real Fender's blue butterflies tend to fly in straighter lines than the ones in this model, do not appear to have any density-dependent behavior, and are in their adult stage only for a short time during the summer. The results that you got from this model certainly apply to many other species, however, and show you the way we would go about deciding what to do for Fender's blue butterflies.

References

Schultz, C. B. 1995. Status of the Fender's Blue Butterfly in Lane County, Oregon: A year of declines. *Report to U.S. Fish and Wildlife Service and the Oregon Natural Heritage Program.*

Biological Pest Control

Approximate time to complete: 3–5 hours

Background

Since people began farming, we have been competing with other creatures for the food we grow. Viral and bacterial diseases sweep through our crops, fungus grows on the plants making them less productive, and insects, birds, rodents, and plenty of other animals are happy to eat the food that we want to harvest for ourselves. Throughout most of history, there was little that could be done to directly combat these creatures, so people relied on natural mechanisms to control the populations of the pests. Natural mechanisms depend on biological interactions, such as predatory insects eating the pest insects, diseases killing the pests, crop plants defending themselves against plant diseases, and so forth. In this century, especially since World War II, we have discovered a whole set of chemicals that will kill pest species. These chemical pesticides (along with commercial fertilizers) have led to a revolution in the way that crops are grown.

Pesticides have problems, however. First of all, most pesticides are not only poisonous to the pest species, they are poisonous to other species as well, sometimes including the humans who apply the pesticides and eat the food that's harvested. Pesticides are also expensive to manufacture and apply. But perhaps the biggest problem is that pest species are very good at evolving resistance to pesticides. There are no known pesticides that are both safe for humans and for which the targets have not evolved at least some resistance. This means that in order to keep the pesticide effective in killing the pests on your crops, every year you need to use more pesticide than the year before, and eventually you must switch to a new pesticide (if there's one available). The people who make pesticides may not mind this too much, but for a farmer, and for the rest of us who pay at the supermarket for the increasing cost of the pesticide, this is hardly amusing.

Because of these problems, people have started to look again at various biological ways of controlling pests. It is much easier for an insect to evolve resistance to a pesticide than it is for that insect to evolve some way of avoiding a hungry spider or a nasty disease. Furthermore, spiders and diseases will evolve right along with the pests in an evolutionary "arms race," so they may not go obsolete. Consequently, farmers and scientists have recently started giving increased attention to the possibility of managing pests biologically. Using other creatures to control pests is known as *biocontrol*. We have some advantages in this approach now that weren't available 100 years ago. First, ecologists understand more about the interactions between predators and prey than they used to (although we still have a long way to go). Second, in the last 20 years we have developed a whole biotechnology industry that is becoming proficient at engineering species to suit our purposes. So we can try to determine ecologically what would be the best type of creature to control our pest, and then we can either go out and look for such a creature, or we can get a biotech company to try and manufacture one.

For the most part, biotechnology is still limited to experimenting with bacteria and plants, and even using these, the technology is in its infancy. So a biotechnology company might consider trying to create a disease that would attack a pest, but could not engineer a spider or another insect to attack the pest. Still, in this lab we're going to pretend that technology has moved ahead significantly, and that we can actually manipulate spiders a bit as well. We're going to take a field that is growing a crop and has a pest insect eating the crop, and we'll try to design the best spider that we can to control the pest in this field. You'll be able to alter three parts of the spider—its eyes, its legs, and its reproduction. You will try to design a species of spider with the optimal combination of eyesight, speed, and reproduction to control the pest.

Outline of This Lab

You work for the biotechnology company called Frankenstein, Inc., specializing in agricultural pest control. You are a member of the spider division, makers of the Paine Killer spider. Your department has spent a lot of money studying this spider, and can now manipulate its genes in a variety of ways.

One set of genes that you can change affects how well the spiders can see. Named the Four-Eyes Cluster by the nerds over in engineering, these genes can be mutated to make keen-eyed spiders that can spot prey a whole 10 meters away, or blind spiders that can't see past their own antennae.

Another set of genes that your department has discovered are affectionately known as the Eight-Legged Universe genes. These control the size of the spiders leg muscles. Again, you can make spiders that can barely scratch themselves or spiders that can cruise along at 10 meters a night. No matter how fast a spider of this species moves, it will only eat once in a night. However, it can reach prey that is farther away if it can move faster.

Finally, your department has found the Love Boat Complex, a third set of genes that determine the conditions under which a spider will reproduce. This species of spider reproduces entirely asexually (after all, they have a job to do, and you don't want them getting distracted). The spiders have one baby at a time, and they won't have this baby until they build up a certain amount of body fat. When a spider eats enough to gain this amount of fat it has a baby; the parent spider then gives half of its body fat to the baby. We'll measure the amount of body fat at which a spider reproduces in terms of how many pest insects it has to eat (at the rate of one prey per night) to gain that much fat. This variable can be changed from one to six insects. With the setting on one pest insect, for example, the spider reproduces each time it eats; set it to six insects, and the spider must eat six consecutive insects before it is able to reproduce.

The spiders live off of their body fat in between feedings, so there is a trade-off here between fast reproduction and starvation. If you make a spider that reproduces really quickly, then it will never build up much body fat, and if there aren't pest insects around for a few days, this spider will quickly starve to death. On the other hand, a spider that builds up a lot of fat before reproducing won't starve nearly as quickly when there are no insects around, but it will reproduce at a slower rate when there are lots of insects.

The Department of Agriculture has contacted Frankenstein, Inc., and is interested in using Paine Killer spiders to keep down populations of a species of aphid that infests wheat. They tell you that these aphids move up to 2 meters a day in search of food, and can totally decimate a stalk of wheat in a night. One of these aphids will reproduce after eating about four stalks of wheat. They also tell you that these aphids have no aversion to spiders. The Department of Agriculture has given your company several wheat fields, each 60 meters by 40 meters, in which to try different types of spiders to see which one will do best at controlling the aphids.

Genetically engineering spiders is hard work, so the biotech people at your company don't want to just start blindly experimenting with different spiders to see which one will work best. Therefore, they have asked you, as the departmental ecologist, to do some modeling and tell them the type of spider that is most likely to control the aphids. To help you, I have constructed a model of a wheat field, with dimensions of 60 meters by 40 meters, just like the experimental fields. An aphid is randomly blown into this field about once a week. After the wheat has been growing for five weeks, 10 spiders will be added to the model. You will be able to give these spiders any combination of the genes described above.

The Department of Agriculture wants your spiders to control the aphids so that there are never more than 150 aphids in the entire field for even a brief time, and there are never more than 100 aphids in the field for a sustained period of time. They want this condition to persist for at least one year after you add the spiders. Your job will be to systematically change the genetics of the spiders to find the combinations that will keep the aphids below these levels. You will then write up your recommendations and give these to your company's gene jockeys, who will go and design the new killer spider.

Incidentally, even without biotechnology, biologists have to make decisions such as those outlined in this exercise. In particular, entomologists look at the life history traits of existing ladybugs, or spiders, or wasps and ask which species possess the traits most suited to biocontrol. They then often try to mass raise these "beneficials" in the laboratory for later release in fields. So the ideas we'll develop in this lab apply today, even without spiffy new biotechnology advances.

<Optional> For those of you who are interested in the nitty-gritty of the model, here is a more detailed description. Every day, 50 stalks of wheat grow up in random squares around the field. If a stalk of wheat grows up in a square that already contains another stalk of wheat, the first stalk dies (there is a maximum of one stalk per square).

The aphids have the following behavior. Every day, an aphid looks 2 squares in every direction. If it sees any stalks of wheat within that distance, then it moves towards the nearest stalk at a speed of 2 squares/day. If the aphid doesn't see any wheat, it moves in a random direction. Every time an aphid lands on a square with wheat in it, the aphid eats the stalk of wheat in that square and gains 1 unit of energy. If the aphid ever gets above 4 units of energy, it has a baby aphid, and it splits its total energy between itself and the baby (each ends up with 2 units of energy). Every day an aphid uses up 1/5 of a unit of energy, whether or not it eats. This is the energy it must spend just to stay alive. If an aphid gets to 0 energy, it dies.

The spiders are modeled in the same way as the aphids, except instead of stalks of wheat, they search out aphids to eat, and you can change some of their parameters. Also, the spider cost of living is a little lower than that of the aphids (1/7 of a unit of energy per day).

The Lab

1. Run EcoBeaker (double-click on its icon).

2. Open the situation "Pest Control" (use the OPEN command in the File menu).

You should see several windows laid out on the screen as follows:

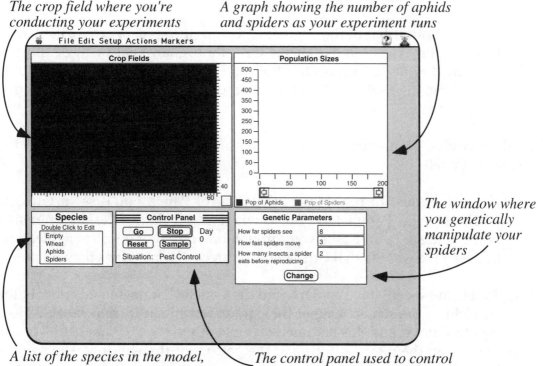

The crop field where you're conducting your experiments

A graph showing the number of aphids and spiders as your experiment runs

The window where you genetically manipulate your spiders

A list of the species in the model, showing each of their colors

The control panel used to control running and stopping the model

The upper-left window shows a view of the field where you'll be doing your experiments. Each creature that appears in that field will be colored as shown in the Species window. The window on the upper right labeled Population Sizes is a graph that plots the number of aphids and spiders as the model runs. Below that is the Genetic Parameters window, which you'll use to make different types of spiders. The window in the middle is the Control Panel, used to control the model.

3. Run the model (press the GO button in the Control Panel) to see what the model looks like.

   For now, focus on the top two windows showing the field of wheat and the populations sizes. You'll see green stalks of wheat shoot up out of the ground at a fixed rate, and every now and then a blue aphid will blow into the field. The aphid will quickly start eating wheat and reproducing. After 35 days have gone by, 10 spiders will be dropped onto the field. These spiders will run around eating the aphids.

   Watch the population graph to see if you are meeting the conditions set out by the Department of Agriculture. The time axis is in days, and you can use the scroll bar at the bottom of the window to scroll the graph backwards if you miss something.

   You will notice that in the beginning, before the spiders arrive, there may be an outbreak of aphids. This is OK—the Department of Agriculture is concerned with outbreaks after the spiders are added and have had time to establish themselves, so you don't need to worry about an outbreak that starts in the first 35 days.

4. When you've seen enough, stop the model (press the STOP button on the Control Panel).

   If you look just to the right of the Control Panel, you'll see a window labeled Genetic Parameters. This is where you'll change the characteristics of the spiders. Let me take you through one round of genetic engineering. Let's say we want to make the spiders move twice as fast as they do now.

5. To double the speed of the spiders, find the Genetic Parameters window in the lower right of the screen. One of the parameters given there is labeled How fast spiders move, and this parameter is currently set to 3. Change the 3 to a 6. Then click on the CHANGE button at the bottom of the window. The spiders will now be able to move 6 squares (6 meters) per day.

6. Reset the model (push the RESET button in the Control Panel) so that everything will start again from the beginning.

7. Run the model (GO).

> You can change how far the spiders can see or how many aphids they need to eat in order to reproduce in the same way as you changed how fast they move, by changing the appropriate number in the Genetics window. Each time you change a number in the Genetics window, don't forget to click on the CHANGE button and then Reset the model.

8. Stop the model (STOP).

9. Before you start seriously exploring different types of spiders to see which one works best, make a prediction about which type of spider you think will be most effective. (In other words, should they be fast or slow, see well or poorly, reproduce quickly or slowly, or somewhere in the middle on all these parameters?) Write this prediction down, so you can look later and see how good your intuition was.

> In order to make your report to the genetic engineers, you will want to systematically change the traits of the spiders. Because the traits interact with each other, changing one of these traits will have different effects if the other two are set to different values. Therefore, you will probably want to change them in pairs.

> Let me suggest one way of summarizing the data you'll collect. You can make a graph with one axis representing one of the traits, and the other axis representing a second trait. You can then fill in each position in the graph with a symbol that indicates whether that combination of traits leads to successful control of the aphids or not. If you pick speed and eyesight as the two axes for a graph, it might look like this:

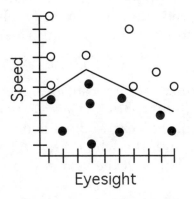

In this graph, open circles show combinations under which the aphids are controlled, and filled circles shows combinations where the aphids aren't controlled.

Note that you don't have to be exhaustive in your modeling—not every point in this kind of graph needs to be filled in for you to get a general idea which combinations of traits are most likely to work. You only need to give the genetics people a general idea of where they should start. Anyway, your model is very simplistic, so even if you wanted to be precise, there is no way you could predict exactly which traits would be best. However, you should probably examine all three of the traits at least once. That is, make at least two graphs like that above.

Remember that there are limits to the values you can give to each trait, as described in the Outline above. Here's a summary.

- Eyesight can range from 0 to 10 meters.

- Speed can range from 0 to 10 meters/day.

- Reproduction can happen after eating 1 to 6 aphids.

Remember that the Department of Agriculture specified that there should be no aphid outbreaks greater than 150 aphids, or sustained levels of aphids above 100, for at least one year after the spiders are added. Finally, note that this model has a lot of randomness in it, so that if you rerun the model with the same parameters, you may get different results the second time.

10. When you are done, write up a short (under two pages, not including graphs) report on your findings for the gene jockeys in your company, and make a recommendation for what type of spider they should try first. If you want to include printouts of the population graph in this report, you can print the graph as follows. Click once anywhere in the graph to make its window active (the title bar of the window should darken). Then select the PRINT command from the File menu.

<Optional> When farmers apply pesticides, they almost never apply them just once in the growing season. In this lab, we only put spiders into the field once at the beginning of the year, and we had to design the spiders so that they could survive through the whole year. What if we could periodically add in more spiders throughout the year, instead of just adding them once?

Would this change the type of spider we would want to use?

In EcoBeaker, we can change the model to add a spider periodically instead of adding a whole bunch all at once. This involves delving a little deeper into the program, and if you get confused you may want to refer to the EcoBeaker manual.

11. Find the Species window, and double-click (click twice in a row) on SPIDERS. A big dialog box will appear. We only need one of the many items here and can ignore the rest. Look on the left side of the dialog box for the label Settlement Proc. Just under that is a box that is currently showing Fixed @ Time X. Click and hold down the mouse button in this box and a menu will pop up. This menu lets you select which settlement procedure is used by EcoBeaker for the spiders. Look for the Fixed settlement procedure and select that. Then let go of the mouse button.

12. Now look just below the pop-up menu to find the SETTLEMENT PARAMS button and click on it.

A second dialog box will appear, and this dialog box will contain a number called the Settlement Rate. This is the number of spiders that are dropped onto the field per day. If we wanted to add one spider a day, we'd make this a 1. If we wanted one spider a week, we would make this a 0.14 (1/7 spiders per day) and then on average, one spider would be dropped into the field per week. There is also a button in this dialog box called Can Settle Over, which specifies what a spider can land on and survive. We'll make a spider be able to survive no matter what it lands on when it's dropped into the field.

13. Click on the CAN SETTLE OVER button. In the dialog box that appears, check all the species. Then click on OK. You won't need to mess with this again.

14. For a start, set the settlement rate to 0.5 so that on average, one spider is dropped into the field every other day. Then click on OK in this dialog box and in the original dialog box. You should now be back to the regular screen.

You can now do the same type of analysis for this model as you did for the original one, except now you have a fourth parameter to play with, namely how often a spider is dropped into the world. Before you start playing, again make a prediction of what type of spider and introduction procedure will work best, and write down this prediction.

15. Are your recommendations different for the type of spider to use if you can drop the spiders into the field more than once a year? Keeping in mind that growing and dropping spiders cost money, what would your overall recommendation be?

More Things to Try

If you want to play with this model more, here's a few things you might try. Spray pesticides along with adding the spiders. This experiment would simulate what actually happens when farmers spray pesticides. You can use the Hurricane procedure to simulate a pesticide (see the manual for a description of how to use this). You might also want to play with this model from the standpoint of predator–prey theory. You could try changing the characteristics of the aphids along with the spiders and see what effect that has. Perhaps more interestingly, you could add in a second prey species or a second predator and see what that does to the dynamics. Or you could add in something that eats spiders. In all these cases, you'll probably want to periodically add in more individuals from these species as described in steps 11–15, since the model is inherently unstable. Does increasing the number of species in the model make it more or less stable?

Notes and Comments

One thing we didn't play with in this lab that might be important to successfully controlling pests is the way in which the fields are arranged. In this lab we grew our wheat in one big field. We could have also tried dividing the field up into many smaller fields, each separated from each other by a barrier that was hard for the aphids and/or spiders to cross. A barrier could be some water, for instance, or fields of some other crop, or some bushes. Growing crops in many small fields instead of a few large ones can help control pests in some situations for a number of reasons. First, when an outbreak of pests occurs, it will sweep through one small field but then be slowed down at the edges. This slowing can give the predators time to catch up (in numbers) and stop the outbreak from proceeding further. Second, the predator–prey cycles will be somewhat different in each field, so that if in one field you have many aphids but few spiders, one of the neighboring fields may have many spiders and few aphids, and some of the excess spiders in the neighboring field may find their way to this field and help stop the aphid outbreak. The idea of using many smaller fields instead of one large one is taken to the extreme in something known as *intercropping*, in which two or more crops are grown in the same field. In an intercropped field, the two crops would be interspersed. Thus if you had wheat and corn, you might plant a stalk of wheat, then a stalk of corn, then another stalk of wheat, then another stalk of corn, and so on. There is active research on all of these ideas right now in a variety of agricultural settings, from small-and medium-sized farms in the United States to traditional rice paddies in Indonesia.

Another way of interrupting the outbreaks of aphids is to rotate the crops planted in a field from year to year. Rotation works because the pests and diseases that affect one type of crop usually won't be as bad for other types. If you grow wheat this year and corn next year, all of this year's wheat pests will go hungry in next year's corn field, and all of next year's corn pests will go hungry in the following year's wheat.

Finally, consider what would happen if we started applying pesticides to this field. Let's say we had a pesticide that killed 99% of all the insects and spiders in the field. On average, a few aphids would probably survive. As you saw in this model, those aphids would then be released from the predation of the spiders, and their populations could explode and do much more damage than if the pesticide had never been used. Even if we had a selective pesticide that killed only aphids, and not spiders, we might still have problems. With almost all the aphids dead, most of the spiders would starve to death (just like the spiders starved to death after eating all the aphids). Again, the few remaining aphids would be released from predation pressure and their population could explode. These phenomena are not just theoretical musings. This type of predatory release actually happens when people use pesticides. The usual response is to reapply the pesticide many times throughout the growing season.

From a more theoretical standpoint, as you were watching the population sizes in this model you probably noticed cycles of predators and prey. The aphid population would grow higher, and this would be followed by growth of the spider population. As the spiders grew more numerous, they would start eating the aphids and bring down the aphid population size. As the number of aphids decreased, the spiders would begin to starve and the spider population would get lower, starting the whole cycle over again. In ecology, we traditionally think of predators and prey interacting in cycles like this, and classical ecological theory such as the Lotka–Volterra predator–prey equations predict these cycles. In real life, you cannot always tell whether the cycles are driven by the predators eating the prey, or whether the prey populations are going up and down for some other reason and the predator population is just following along. Still, cycles like this undoubtedly happen in the natural world as well as in agricultural fields.

I don't want this laboratory to leave you with the impression that pesticides are all useless or evil. Pesticides may enable us to grow more food than we could have without them, and modern agriculture practices on the whole are very successful, as we witness every time we enter a supermarket. However, using biology to control biological pests has the potential to improve our farming in the future, especially as more and more pests evolve resistance to our pesticides.

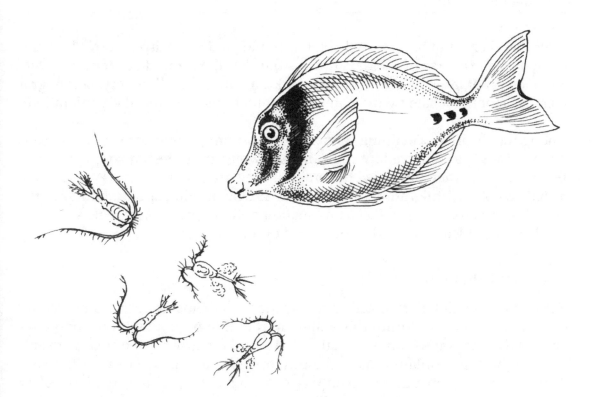

Predator Avoidance in Aquatic Systems

Suggested by Carol Eunmi Lee, University of Washington

Approximate time to complete: 2–4 hours

Background

Aquatic habitats, such as lakes, oceans, and rivers, are filled with lots of little animals collectively called *zooplankton*. Zooplankton are comprised of many species of crustaceans, such as copepods, euphausiids, and cladocerans (the shrimp you get in restaurants are zooplankton). Also included are the gelatinous zooplankton, such as medusae, ctenophores, salps, and doliolids (commonly called jellyfish). Zooplankton are by definition weak swimmers and cannot move very much relative to the horizontal fluid flow. They can, however, move considerable distances up and down, on the order of hundreds of meters per day. Vertical migrations of zooplankton were first observed around 1966 by sonar systems. Sonar systems send out waves of sound and then listen for the echoes from the sound bouncing off of objects in the water. People noticed layers in the water that seemed to bounce back a lot of sound, and these layers moved up and down over the course of the day. It turns out that the "sound-scattering" layers are made up of large numbers of zooplankton. Since this discovery, oceanographers have hotly debated why many zooplankton undergo this daily migration.

Among the theories explaining why zooplankton might migrate is one that says they migrate to avoid predation. In this lab we'll try to design zooplankton in such a way that they minimize the amount they are preyed upon. We'll be able to change not only migration behavior, but also swimming speed and reproductive rates. We'll try to see which combination of these parameters makes the best zooplankton, at least under the conditions of this model.

Outline of this Lab

In this lab you will be presented with a model of a 60-meter-deep water column in an aquatic system. Within this water column live phytoplankton, zooplankton, and fish. As you would expect, the phytoplankton grow in the areas with sufficient light, the zooplankton eat the phytoplankton, and the fish eat the zooplankton. The phytoplankton can only grow within the top 25 meters, since below that there is never enough light for them. Likewise, the fish can only hunt zooplankton within the top 25 meters, since below that its too dark for them to see the zooplankton. The zooplankton in this model are mainly filter-feeders, so they can get food equally well in light or dark.

Each species immigrates into and emigrates from the area of lake on the screen at a low rate when you run the model, and all species also reproduce based on how much food they have available. Phytoplankton are assumed to always have enough light, so when they are low in abundance and there is nothing eating them, their population will grow exponentially. At higher levels they start to run out of nutrients and their growth slows down, so that overall their growth curve is logistic. The zooplankton reproduce based on the amount of food they eat. When a zooplankton has accumulated enough energy through eating phytoplankton, it produces another zooplanktor. Each time it reproduces, it splits its energy with its offspring, so that each now has half the energy level of the parent. Zooplankton also need energy to stay alive and keep swimming, so they can only accumulate energy if they eat more phytoplankton than they need for their metabolic needs. If they don't get enough phytoplankton to satisfy their metabolic needs, they eventually die. The only exception to this is when the zooplankton are sitting still in the dark lower zone of the water, in which case they use no energy. Fish are modeled in the same way as zooplankton, except they need much more energy in order to reproduce since they are bigger.

You are going to play Darwin in this model. Your goal is to design a zooplankton species that can maintain the highest population level in the face of both limited food and predation by the fish. Evolution normally acts upon individuals, of course, but we'll assume that in this case, a large population indicates that each individual is also doing well (note that this is not always a good assumption). You will be able to change the swimming speed of the zooplankton, the amount of energy a zooplankton needs to accumulate before reproducing, and the timing of the zooplankton daily vertical migration. You will be able to change all of these parameters within reasonable limits in a quest to design the optimal zooplankton for this situation.

The Lab

1. Run EcoBeaker (double-click on its icon).

2. Open the situation "Copepods" (use the OPEN command in the File menu).

You should see several windows laid out on the screen as follows:

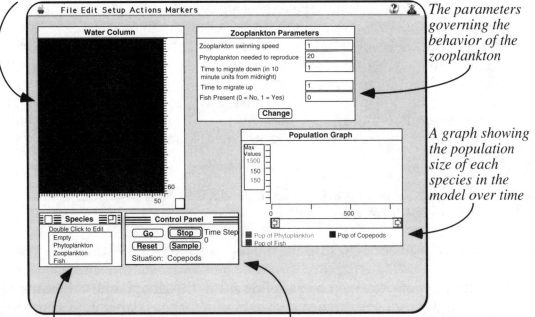

A vertical slice through a body of water where we are observing zooplankton and fish

The parameters governing the behavior of the zooplankton

A graph showing the population size of each species in the model over time

A list of the species in the model, showing each of their colors

The control panel used to control running and stopping the model

The main window in the upper left shows a slice through the Water Column in some lake or ocean. When you run the model, you'll see creatures swimming around in this area. The abundance of creatures of each species is shown in the Population Graph in the lower right. There are many more phytoplankton than zooplankton or fish, so the three species are plotted with different scales. The top of the graph corresponds to 1500 phytoplankton (plotted in green), but only 150 zooplankton (in blue) or fish (in red).

At the bottom of the screen is a window that shows you the colors of each species in the model, another window called the Control Panel that lets you start and stop the model. The Control Panel also shows how much time has gone by since the model was started. Notice the time unit. Each unit of time in the model is 10 minutes, so that 6 units of time equals 1 hour. To help you convert this weird time unit, I give a conversion chart below.

The fifth window on the screen, labeled Zooplankton Parameters, is where you will design your zooplankton, using a method explained below.

3. Run the model (push the GO button in the Control Panel).

In the Water Column you will see green phytoplankton reproducing in the upper layer of the water column, in the euphotic (light-lit) zone. As each day goes by, the level of light in the euphotic zone will change from bright daylight to dark nighttime, with low-light dusk and dawn in between the day and night. There is never any light below the euphotic zone. You will also see blue zooplankton grazing on the phytoplankton. In the Zooplankton Parameters box, the Fish Present parameter should equal 0. That means there are no fish present. When you add fish in later, they will be red.

4. Observe the graph for about 15 to 20 days (3600 to 4800 time steps). What are the abundance patterns over time? Explain.

5. When you've seen enough, stop the model (push the STOP button).

6. Add fish by typing "1" next to Fish Present in the Zooplankton Parameters window. You must click on the CHANGE button for the change to become registered. Now run the model again (GO). What happens to the phytoplankton and zooplankton abundances when you add fish? Why?

Your task is to maximize the abundance of zooplankton. You have several choices:

(a) *Vary zooplankton swimming speed.* The zooplankton swimming speed is currently set at 1 m/10 minutes. You can increase this up to 20 m/10 min. Changing swimming speed will change the encounter rates with food and predators.

(b) *Vary zooplankton reproduction rate,* by changing the number of phytoplankton needed to reproduce. The reproduction rate is currently set at 20 phytoplankton needed to make one new zooplankton. Lower numbers would increase the reproduction rate. You can set the number of phytoplankton needed to reproduce anywhere from 1 to 100 phytoplankton.

In this model, there is a cost to a zooplankton in increasing its reproductive rate because a zooplankton gives up half its energy to its offspring when it reproduces. This means that it will be more likely to die from starvation afterwards if there is not enough phytoplankton around. So increasing the reproductive rate also increases how prone the zooplankton are to starvation.

(c) *Vary timing of vertical migration.* The time you indicate is the time the zooplankton begin migrating up or down. Setting the Time to go up and Time to go down equal to each other will eliminate migration. Migrating allows the zooplankton to move in and out of the surface layer, where both food and predators are.

Time in this model is displayed in ten minute increments beginning at midnight. You must multiply hours by 6 to get your desired time setting.

For instance,

Time		Setting
12 midnight	=	0
6 A.M.	=	36
12 noon	=	72
6 P.M.	=	108

7. After you make each modification, click on CHANGE, and then GO.

> You probably want to change variables one at a time. This approach will make it easier for you to see what effect each variable has. You can then systematically change two variables relative to each other.

> Remember to click on the CHANGE button each time you change a parameter, or the new value will not be registered in the model.

8. In the face of fish predation, what conditions allow the zooplankton population to reach a maximum abundance? Explain.

9. Try to generalize your results from this model. Can you come up with a hypothesis for when it might be advantageous for zooplankton to be diurnal (come up at day), versus nocturnal (come up at night), and when it might be better for them not to migrate at all?

Notes and Comments

The cause of vertical migration behavior in zooplankton has been a source of controversy ever since it was discovered, inspiring many hypotheses and studies. In aquatic habitats, some species in nearly every animal phylum exhibit diel (twice a day) vertical migrations, even larval stages of some benthic (bottom-dwelling) species. If you go to a particular location that has vertically migrating zooplankton, though, you will also be able to find other species that don't migrate. Even within one species, the pattern of migration can vary with age or developmental stage, with sex, and among individuals of the same sex and stage.

Among the hypotheses put forth about vertical migration are: (1) it is driven by an energetic trade-off related to water temperature, (2) it is a means to avoid predation, and (3) it helps a zooplankton stay in a location with high water flow. The energy trade-off hypothesis argues that zooplankton use up less energy when they are in cold water than when they are in warm water. Phytoplankton grow in the upper part of the water column, which will be warmer than the lower layers of water in many places. Thus a zooplankton might be able to conserve energy by feeding in the warm upper waters, then swimming to the cooler lower waters where its metabolism will be lower to digest what it has eaten.

The predator–avoidance hypothesis was what we explored here, and much evidence suggests this is why many species vertically migrate. For instance, there are species of zooplankton that can be put into a big enclosure without fish, and they won't vertically migrate. However, if fish are put into the enclosure, the zooplankton will almost immediately start vertically migrating. Other zooplankton are genetically programmed to vertically migrate, but you can again show that this programmed behavior was caused by selective pressure of predators. If you didn't have genes that told you to vertically migrate, you got eaten. A final piece of evidence that migration is driven by predation is the evidence of *reverse migration* in some species. This reversal occurred when a large species of zooplankton migrated to the surface during the night and returned to deeper water during the day to avoid visual predation, as our species did in this lab. If that large species ate a smaller species of zooplankton, then the smaller species might migrate the opposite direction, to the surface during the day and to depth at night, in an effort to avoid the larger zooplankton species.

Another plausible hypothesis why some zooplankton migrate is to stay where they are. If a zooplanktor lives in a place where the water flow is high, such as an estuary, and if the zooplanktor stayed in one part of the water column it would get washed out to sea. An individual can prevent this by adjusting its position throughout the day so that sometimes it's in a part of the water column flowing one way, and other times it's in water flowing the other way, keeping its average position constant.

In a broader context, vertical migration is a clear example of how the behavior of animals relates to their environment and to the other species with which they interact. Just like other defining characteristics (such as size, shape, color, and so on), the behavior of an organism evolves and is shaped by natural selection. Behavioral differences between species are just as important as other differences, and can have just as large an effect on the ecosystems in which they live. Thus the field of behavioral ecology has made important contributions to our understanding of the world around us.

Below I've included a few of the many references on vertical migration. These will lead you to more literature if you are interested in exploring further.

References

Barham, E. G. 1966. Deep scattering layer migration and composition: observations from a diving saucer. *Science*. 151:1399–1402.

Bollens, S. B. and B. W. Frost. 1989. Predator-induced diel vertical migration in a planktonic copepod. *Journal of Plankton Research*. 11:1047–1065.

Cronin, T. W. and R. B. Forward, Jr. 1979. Tidal vertical migration: an endogenous rhythm in estuarine crab larvae. *Science*. 205:1020–1022.

McLaren, I. 1974. Demographic strategy of vertical migration by a marine copepod. *American Naturalist.* 108:91–102.

Neill, W. E. 1990. Induced vertical migration in copepods as a defence against invertebrate predation. *Nature.* 345:524–526.

Watt, P. J. and S. Young. 1992. Genetic control of predator avoidance behavior in *Daphnia. Freshwater Biology.* 28:363–367.

Quadrat Sampling

Suggested by Amatzia Genin and Emmanuel Noy-Meir, Hebrew University

Approximate time to complete: 2–3 hours

Background

In ecology you frequently encounter situations where you want to know how how many individuals of a certain species populate a particular area, but there are too many individuals or the area is too big to actually count every last creature. To solve this problem we can pick a small piece of the area, count what's in that small piece, then multiply by the total area to get an estimate of the number of creatures in the total area. This is known as *sampling using quadrats*, and is perhaps the most common sampling technique used by ecologists. When counting small things, the quadrats are usually delineated by some rectangular frame cobbled together using whatever materials are handy (strips of wood, pvc pipe), which you can then lay down on the ground to show the area in which to count critters. For bigger things you might lay out quadrats using poles and string, or use some more fancy technique.

Since even sampling a small part of the total population usually takes quite a bit of time, we'd like to know the minimum number of samples we need to take in order to get an accurate estimate of the population size. We would also like to know if the way we build and arrange our quadrats makes a difference in how accurate our sampling technique is. For instance, should we use one big quadrat or many small quadrats? Should we lay out our quadrats randomly or should we arrange them systematically around the study site? Should each quadrat be square or rectangular?

An additional factor to consider is that sampling techniques that work well for one type of population may not work so well for another. You might want to use a different sampling scheme for a species whose members are scattered randomly about the landscape than you would for another species that's found in patches.

In this lab I've made a landscape with three different species, each of which is distributed in a different way. We'll try sampling these three species in some of the different ways mentioned above. The goal is to figure out the most efficient sampling scheme where we can get a reasonably accurate estimate of the population size.

Outline of This Lab

I have made a 50-meter by 50-meter landscape with three different types of plants on it, each a different color. The green plants are growing in completely random places around the landscape. The blue plants grow much better when they are near other blue plants, so they are clumped together (a patchy distribution). The red plants compete strongly with each other, so they do not like to grow nearby other red plants. This makes the red plants space themselves out in a somewhat even pattern. There are exactly 80 individuals of each plant in the whole area.

You will try sampling this area using quadrats in a variety of ways. The goal is to get the best possible estimate of each species population size using the minimum amount of sampling effort. Sampling effort in this case is how much area you sampled. If your quadrats are 5 m × 5 m, and you use four of them, then your sampling effort consisted of counting everything in a total of $5 \times 5 \times 4 = 100$ m^2 of land.

You will be sampling from a landscape that has a total area of 50 m × 50 m = 2500 m^2. This means that to get a population estimate from one of your samples, you will have to divide the number of individuals counted by the area sampled, and then multiply by the 2500 m^2 area of the whole landscape. For instance, if you found 25 red plants in the four 5 m × 5 m quadrat sampling scheme described above, you would divide the 25 plants by the 100 m² area sampled to get a density of 0.25 plants per m², then multiply by the total 2500 m² of the whole landscape to get an estimated population of 250 red plants. In equation form, this looks like:

$$\frac{N}{A_{Sampled}} \times A_{Total}$$

where N is the number of plants you counted, $A_{Sampled}$ is the area that you sampled, and A_{Total} is the total area you are sampling from, in this case 2500 m².

We will start by asking how large an area we need to sample to get good population estimates of each species if we only use a single, randomly placed quadrat. We'll then try a small quadrat, and see how many small quadrats we need to use in order to get good population estimates. We'll also look at whether the number of quadrats we need changes if we lay them out randomly or systematically. Finally, we'll look at how changing the shape of the quadrats changes how good our estimates are.

Since "good" is a relative word, you will have to decide what you mean by a good sampling scheme for estimating the population sizes. In order to make this decision, you will need to take more than one sample with each sampling scheme. For instance, if you were trying to sample with one 5 m × 5 m quadrat, and you only did it once, you might just by chance get a very good (or very bad) estimate of the population sizes of all three species. Since you only tried that sampling scheme once, you wouldn't know if you were just lucky to get the correct numbers, or if you could expect a single 5 m × 5 m quadrat to consistently give you good estimates. Thus, you should try using a single 5 m × 5 m quadrat several times, and see whether you get a good estimate each time. You will still have to decide what your criteria is for "good estimate each time," as well as how many times to try each scheme. I would suggest trying each scheme at least three times.

Note: This lab can be short or long, depending on how many different parameters of the sampling technique you try varying. If you are doing this in a class, you might split up the work, so that different groups try different sampling schemes, then compare results at the end.

The Lab

1. Run EcoBeaker (double-click on its icon).

2. Open the situation "Random Sampling" (use the OPEN command in the File menu).

After the situation loads, you should see several windows laid out on the screen as follows:

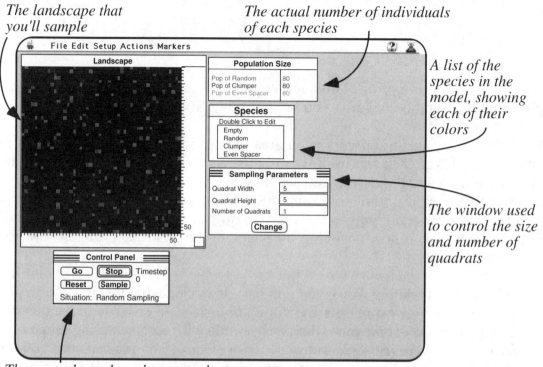

The landscape that you'll sample

The actual number of individuals of each species

A list of the species in the model, showing each of their colors

The window used to control the size and number of quadrats

The control panel used to control running and stopping the model

The main window in the upper left of the screen shows the Landscape that we'll be sampling. To the right of this main window is the Species window, which identifies the color of each of the three species. Above the species window is a small table showing the number of individuals of each species in the landscape, currently 80 of each. We will not change any aspect of the landscape throughout this lab, but when you finish you may want to try increasing densities or changing the arrangement of species to see what effects these changes have (see "More Things to Try").

The final two windows are the main ones you'll pay attention to. One of them is labeled Sampling Parameters, and lets you change the width, height, and number of quadrats which you use to sample the populations. The other is the Control Panel, which will let you actually take a sample.

As a start, let's try using a single sampling quadrat, changing its size, and seeing how large we need to make it in order to effectively sample the population.

3. Click on the SAMPLE button in the Control Panel to take a sample. A square will be drawn on the Landscape window, showing the area that is being sampled. Just below the Landscape window a dialog box will appear, telling you how many individuals of each species were found within this square. Write down both the number of individuals, and the width and height of the area you sampled (which is currently 5 m × 5 m, as shown in the Sampling Parameters window).

4. Repeat step 3 several more times, so that you get range of different estimates that tell you whether you can expect this scheme to consistently give you good results.

5. Now increase the size of the sampling quadrat to a larger size such as 10 m × 10 m. To do this, go to the Sampling Parameters window and type in "10" for both the Quadrat Width and Quadrat Height items. Then click on the CHANGE button in the Sampling Parameters window to make the change in size you specified go into effect.

6. Repeat steps 3 and 4 with the enlarged quadrat.

7. Continue increasing the quadrat size in some regular way, keeping the shape of the quadrats square, and taking several samples at each quadrat size. Remember to write down both the quadrat sizes and the number of individuals counted for each sample.

8. When you have finished your sequence, estimate the true population size from each sample you took. Then look at these estimates.

> One nice way to look at them is to make a plot of population estimate versus quadrat area. This will show you visually what happened to your estimates as you increased the quadrat size. You will have several points plotted for each quadrat size, from each of the samples you took at that size, and this will show you the range of accuracy you might expect from a quadrat of that size. Make sure to plot each species separately.

9. What are the shape of your "estimated population sizes versus quadrat area" graphs for each species? Is there some point where increasing the size of the quadrat seems to make little difference in accuracy? Is this point the same for all the species?

> I will now describe how you can try a variety of other sampling schemes. For each one, try to make a graph similar to the one you made above for different

quadrat sizes. The next manipulation I'll describe involves using small quadrats, and increasing the number of quadrats used for each sample.

10. Set the quadrat width and height back to 5 m each. Take a set of samples as you did in steps 3 and 4. Write down the results from each sample, and also note that you only used a single quadrat.

11. Now go to the Sampling Parameters window and change the Number of Quadrats to 2. Take several samples, as above.

This time, when you take a sample, first one 5 m × 5 m quadrat will appear, as before, and a dialog box will tell you how many individuals of each species were found in that quadrat. When you click on the Continue button, a second 5 m × 5 m quadrat will be laid down, and again you'll receive a report of how many individuals were found in the second quadrat. Finally, a third dialog box will appear giving you the total number of individuals found within both quadrats. For the purposes of this lab, you can simply write down the totals.

Note: I have set up the quadrat sampling so that one quadrat may overlap with another when you are using more than one quadrat per sample. If you decide that you would like only nonoverlapping samples, you can do this as follows. Go to the Setup menu and select the SAMPLING... item. A dialog box will appear. Find the Advanced Stuff button near the upper right of this dialog box, and click on it. A new dialog box will appear. At the top of this second dialog box are three radio buttons under the label Verbosity. Click on the radio button labeled Interactive. Then click on OK in all the dialog boxes.

If you perform the above procedure, from now on when you take a sample you will be asked whether you like the position of each quadrat. You can check on the Landscape window to see whether the quadrat overlays a previously sampled quadrat, and if so, tell EcoBeaker NO, you don't accept that quadrat position. EcoBeaker will then move the quadrat to another random position and ask you again whether that new position is good. You can repeat this indefinitely until the quadrat ends up in a place where it doesn't overlap any others, and then choose YES to accept its position.

12. Continue increasing the number of quadrats you use and sampling several times with each number of quadrats.

13. When you have tried a range of quadrat numbers, make graphs of estimated population sizes versus sampling effort, as you did in steps 8 and 9. Do these

graphs look about the same as those you got with a single quadrat? Does it take more or less sampling effort to get good population estimates with multiple small quadrats than with a single large quadrat? Is the answer different for the different species?

Next you might want to try changing the shape of the quadrats, still keeping their total area constant. For instance, instead of a 10 m × 10 m square quadrat, you might try a 5 m × 20 m quadrat, or a 4 m × 25 m quadrat, or a 2 m × 50 m quadrat, and so on. You could try picking several of the sampling schemes you tried before (either single quadrats or multiple quadrats), and vary the shape of the quadrats to see how that affects your population estimates. You may want to pick schemes you tried before that didn't work as well as you'd like them to, and see if changing the shape of the quadrats makes them work better.

14. Pick one or more of the sampling schemes you used before, and try changing the shape of the quadrats you use by making the Quadrat Width and Quadrat Height different sizes. Again, make graphs of your results and compare these graphs both to each other and to your previous results.

Finally, let's try using a systematic sampling scheme instead of placing quadrats randomly.

15. To change from random quadrats to systematically placed quadrats, find the Setup menu and select the SAMPLING item. A dialog box will appear that lets you change the sampling technique you use, and the parameters of that technique. At the top of the dialog box is a pop-up menu where you can select the sampling technique. Click and hold down the mouse button where it says Random Quadrat. A menu will pop up. From this menu, select SYSTEMATIC.

You will see below the pop-up menu that the parameters for the sampling technique will also change. There are now four parameters. The first two are the same as before, the width and height of the quadrats. The next two parameters specify the distances from one quadrat to the next in the horizontal and vertical directions on the screen, respectively. For instance, if you have 5 m × 5 m quadrats and you specify 10 m between each quadrat horizontally, then a quadrat will be placed in the upper-left corner, another quadrat will be placed starting from 15 to 20 meters (10 meters past the end of the first quadrat), and so on across the screen. If you also specified 10 m between quadrats vertically, then this pattern will be repeated at 15 to 20 meters down from the top, and so on to the bottom of the Landscape.

Unfortunately, you will not be able to change the horizontal and vertical distances between quadrats from the Sampling Parameters window. To change these, you will have to go back to the dialog box by selecting SAMPLING from the Setup menu as above. You can still change the width and height of the quadrats from the Sampling Parameters window as before.

16. To get a feel for how the systematic sampling technique works in EcoBeaker, set the quadrat size to 5 m × 5 m, and put 20 m between neighboring quadrats both horizontally and vertically. Then click OK in the dialog box. Now take a sample.

    As before, you will be shown the number of individuals of each species found for each quadrat sampled, then at the end you'll be given the total number found in all the quadrats.

    With the systematic sampling technique, you can try many of the same exercises as before. For instance, you might try keeping the quadrat size the same, but decreasing the distance between quadrats (thus increasing the number of quadrats used and the total sampling effort). Or you might try increasing the quadrat size while proportionately decreasing the distance between quadrats (so that the number of quadrats used stays the same, but the total sampling effort goes up). Keep track of your changes and make graphs as you did before. You might also try using the same number and size of quadrats as you did in the technique using multiple, randomly placed quadrats, and see whether systematically or randomly placing quadrats work better for a given sampling effort.

17. When you have gotten thoroughly sick of taking samples, sit back and look at all your graphs. If possible, compare your results with the results of other groups that have been doing the same lab. Can you draw any general conclusions? Do some sampling techniques seem to work better than others? Does the best sampling technique change depending on the distribution of organisms on the Landscape (randomly placed versus patchy versus evenly spread out)?

More Things to Try

In this lab we tried sampling one set of distributions, and one population size. If you want to explore further, you might want to change either the distribution patterns or the size of the populations. I used a model to generate the distributions you see, and below I explain how this model works. If you want to, you can change parameters of the model, rerun it, and then try your sampling schemes again and see how well they fare. If you don't want to figure out exactly how the model works in detail, you can change the distributions somewhat simply by running the model a little more. To do that, press the GO button in the Control Panel. When you reach a distribution that you like, press the STOP button.

The way the model works is as follows. All three species use a settlement technique called "Interact". This technique settles a certain number of new individuals onto the Landscape each time step, but the chance of a new individual successfully settling depends on how close the nearest neighbor of the same species is. For the plant that settles in clumps, the closer the nearest neighbor is to a newly settling individual, the more chance that the new individual will successfully take root and survive. For the plant that settles in an even distribution, the farther away the nearest neighbor is to a newly settling individual, the greater the chance that the new individual will survive. And for the randomly distributed species, it doesn't matter where the nearest neighbors are. In addition to settlement, the model includes a chance of death for each already settled individual.

Open up the Species Setup box for the clumped species to see how this settlement technique works in practice. To open the setup box, go to the Species window, and double-click on the name CLUMPED. A fairly complicated dialog box will appear. (For more information on each of the items, look at the EcoBeaker manual). Here we just need to worry about two of the items. If you look in the middle left-hand side of the box, you'll see a label Settlement Procedure. Under that will be the name of the procedure being used, which is the Interact procedure described above, and just under that is a button that says "Settlement Params". Click on the button. This will bring up a second dialog box where we can set the parameters for the Interact settlement procedure.

The parameters work as follows. Number To Settle new individuals are settled in random locations around the Landscape each time step. For each of these individuals, the program looks around the place where they settled to find their nearest neighbor of the same species. If the neighbor is 0 squares away (in other words, if they land on top of a conspecific), then the chance that the new individual will survive is given by Chance @ 0 Dist. If the nearest conspecific is Max Effect Dis" squares away or farther, then the chance that the new individual will survive is given by Chance @ Max Dist. In between 0 and max distance, the chance of successfully settling is in between the chance at 0 distance and the chance at max distance. (It's a linear interpolation).

Exit the settlement parameters dialog box by pushing CANCEL. This will bring you back to the species setup box. Aside from the settlement procedure, the other interesting thing that happens is that individuals die. The chance that any individual will die in any given time step is given in the Transition Matrix, which is show in the right half of the dialog box. Look at the box labeled No Change, which is given the value 0.97. This means that there is a 97% chance that an individual of this species will survive to the next time step. Then look at the box labeled Empty, which has a value of 0.03. This means a 3% chance of death for individuals of this species per time step.

That's the whole model. You can change any of these parameters in any of the species to get new distributions. Have fun.

Notes and Comments

This lab gives you a feel for the types of issues involved in designing a sampling scheme. There are, of course, practical issues as well. It's very easy for a computer to pick random positions in which to set quadrats in an imaginary world on the screen, but in a real field situation, it may be much easier to lay down quadrats systematically. It may also be easier to sample square quadrats than long rectangular ones, or vice-versa. Thus your sampling effort depends not only on the area you sample, but also the technique you choose. Depending on how hard it is to collect samples, you may not have the time or money to do as many sampling exercies as this lab indicates you need for confidence in your results. These and many other issues make designing sampling techniques a potentially tricky business, but a very important one if we want to trust our results.

References

There are many papers and books on sampling techniques for ecology and I can't hope to provide a comprehensive list, but here is one book that is quite good and can lead you to further sources.

Krebs, C. J. 1989. *Ecological Methodology*. Harper and Row Publishers, New York, NY.

Mark and Recapture

Approximate time to complete: 1–2 hours

Background

Sampling a population of animals that moves around a lot can be really tricky. Sampling techniques such as quadrat sampling or line transects won't work if the animals won't stay put while you're trying to count them. This problem is compounded if the density of animals is low, so that chances are you won't even see any of the animals when you're trying to count them. So ecologists have tried to figure out other sampling techniques to determine population sizes for mobile animals—techniques that don't require finding all the animals in a given area at a given time. One of the more widely used techniques is called *mark and recapture sampling*. The idea is that you catch a bunch of individuals of the species you are interested in, mark them somehow, then let them go to mix back into the rest of the population. Then, at some time in the future, you again catch a bunch of individuals and you look to see how many of the individuals you recapture are marked, and how many are not marked. The ratio of the marked to unmarked individuals gives you an estimate of the population size of the whole population.

To see intuitively why this works, consider the following scenario. Let's say you want to estimate the number of frogs in a pond. Let's say (although you don't know this yet) that this pond contains 1000 frogs. To find out the population size, you catch 100 frogs and put a red spot on the back of each one. Then you release these marked frogs back into the pond. You come back the next night, and catch another 100 frogs. You should get around ten marked frogs, and 90 unmarked ones, because one out of every ten frogs is marked (you marked 100 out of 1000 frogs). In equations, you expect:

$$\frac{Num\,Marked}{Total\,Pop\,Size} = \frac{Num\,Marked : Re\,captured}{Num\,Re\,captured}$$

where *Num Marked* is the number of frogs you marked on the first night, *Total Pop Size* is the total number of frogs in the pond, *Num Marked:Recaptured* is the number of frogs caught on the second night that had little red dots on them, and *Num Recaptured* is the total number of frogs captured the second night.

If on the second night you caught nine marked frogs and 91 unmarked frogs, you could estimate the total population of frogs by using the formula:

$$Est.Total\,Pop\,Size = Num\,Marked \times \frac{Num : Re\,captured}{Num\,Marked : Re\,captured}$$

This formula is just a rearrangement of the first formula. The *Total Pop Size* is now an estimate, since it's coming from your sampling. In the above example, that would come out as:

$$Est.Total\,Pop\,Size = 100 \times 100 / 9$$

which equals 1111, not too bad an estimate, at that.

In this lab we'll try out the mark and recapture sampling technique on a population of pigeons in Central Park. We'll play with changing the total number of animals, the number of animals that you mark, and the time you wait before recapturing animals. We'll try to see how each of these affects the accuracy of the technique, and we'll also see that the technique is perhaps a little trickier to get good estimates out of than other sampling techniques.

Outline of This Lab

In this laboratory, you are a brave explorer, going into the wilds of New York City to study that most fearsome of animals, the ever-present, ravenous, unflappable pigeon. While taking care to protect your head from the bombs they drop on unwary adventurers, you capture some pigeons in Central Park, mark them with a bright red color, and then release them to continue their capering quest of consummation. After you have recovered, you come back and stoically stand in Central Park again, snatching any pigeons that are unlucky enough to come your way, noting for posterity whether or not they are marked red. Not content to torture yourself too lightly, you yearn to repeat this pigeon capturing prescription several times, each one a slight variation on the other, until your pigeon counting recipe is beyond reproach.

The Lab

1. Run EcoBeaker (double-click on its icon).

2. Open the situation Pigeons (use the OPEN command in the File menu).

After the situation loads, you should see several windows laid out on the screen as follows:

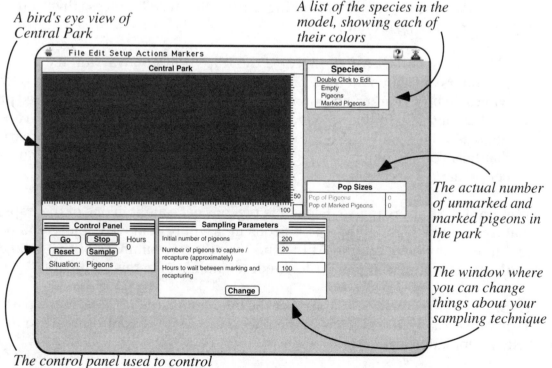

A bird's eye view of Central Park

A list of the species in the model, showing each of their colors

The actual number of unmarked and marked pigeons in the park

The window where you can change things about your sampling technique

The control panel used to control running and stopping the model

The big window in the top left is a view of Central Park, NYC (minus the smog and pedestrians). Next to it is a window called Species, showing the colors of marked and unmarked pigeons; below that is a window called Pop Sizes giving the number of unmarked and marked pigeons in the park. At the bottom of the screen is the Control Panel, which you'll use to run the model and take samples, and the Sampling Parameters window, which lets you change the way you take your samples as described below.

3. Start the simulation running (push the GO button). You will see a bunch of pigeons appear on the screen and start flying around randomly. (OK, they're square pigeons—you have a problem with that?).

4. When you've had enough fun watching the birds fly around, click on the SAM-PLE button in the Control Panel to start marking pigeons. Don't stop the model from running before doing this; if you did, start it again (GO).

    You will see four traps appear in Central Park. Any pigeons that wander into these traps will be caught and you'll have the opportunity to mark it red, until the number of red pigeons is more than 20.

5. Each day, any trap that caught some pigeons will be highlighted and you'll be asked if you want to mark those pigeons. Click on the YES button to mark them. When you have more than 20 marked pigeons, the traps will disappear. (Make sure the model is running during this whole process this so that pigeons are moving around).

    After marking at least 20 pigeons, the traps will go away for 100 hours (the hours are shown in the Control Panel). When the traps reappear, they will catch both marked and unmarked pigeons until they have captured at least 20 pigeons. In this second round of trapping, each pigeon that's captured will be removed from the park. At the end, you'll get a summary that tells you the number of pigeons you originally marked, and the number of marked and unmarked pigeons you recaptured.

6. Wait until the traps reappear. Each time pigeons get caught in one of the traps, that trap will be highlighted, the pigeons in it will be removed, and you'll be told how many were caught. At the end, you'll get a summary of that whole mark and recapture sample. Write down the numbers in this summary.

7. From the sample, make an estimate of the total number of pigeons in the park. There are actually 200 pigeons in the park. How close is your estimate to the actual population size?

8. Reset the model (press the RESET button in the Control Panel) and repeat steps 4 through 7 to take another sample. How close was your estimate this time? How much difference was there between your first and second estimates? How consistent is this sampling technique?

> The rest of this lab lets you play with the number of pigeons in the park, the number of pigeons that you mark and recapture, and the time you wait between marking and recapturing. Your goal is to see how each of these parameters affects the accuracy of your population estimate.

> To change any of these parameters, look for the Sampling Parameters window underneath the Central Park window. The top number in this window lets you set the number of pigeons flying around the park. The middle number lets you set how many pigeons to catch and mark (approximately), and how many pigeons to recapture. The bottom number is how many hours you wait between marking and recapturing. Each time you change one of these numbers, you should then reset the model (click RESET on the Control Panel).

> To get you started, let me show you how to decrease the time you wait between capture and recapture to 50 hours.

9. To wait 50 hours between the time you finish marking pigeons and when you start recapturing, change the "Hours to wait between marking and recapturing" in the Sampling Parameters window to 50. Then click RESET on the Control Panel. Now start the model running again (GO), and take a sample (SAMPLE).

10. Repeat steps 5–8.

> OK, the rest is up to you. Try playing with the different parameters until your curiosity is satisfied. You will probably want to start by changing paramenters one at a time.

11. Can you draw any conclusions about this sampling technique in general? How accurate are the results, and what seems to affect their accuracy the most?

Notes and Comments

You may have noticed that the population estimates you got from this method were not too accurate. Mark and recapture is a notoriously hard technique to use and interpret, and it's not easy to get good estimates from it. One problem you might have noticed is that if you don't wait long enough between capturing and recapturing birds, the marked birds don't have a chance to completely mix back into the population. You then recapture more marked birds than you should for a randomly mixed population, and have too low a population estimate. This will also be a problem for animals that don't move around randomly, since there may never be complete mixing.

In addition to the problems you may have had here, mark and recapture is confounded if animals are born or die while you are doing your capturing and recapturing. The simplest methods of analyzing mark–recapture data also must assume that there are no animals moving into or out of the population from other populations. In addition, you must assume that marked and unmarked individuals are equally likely to be caught, equally likely to die, and that whatever mark you use can't be washed off or lost some other way. Can you see why these would be problems? Sometimes you can use some fairly sophisticated statistics to get around these problems. More sophisticated analyses can even estimate birth and death rates from the mark–recapture sampling data, by recapturing several times. Overall, you can see this is not the cleanest sampling technique devised. Nevertheless, it's one of the only techniques we can use for mobile animals, so it has been quite important in many ecological studies.

References

Krebs, C. J. 1989. *Ecological Methodology*. Harper and Row Publishers, New York, NY.

Breeding Bird Sampling Design

Approximate time to complete: 3–5 hours

Background

In the news these days we often read warnings about some threatening environmental problem—deforestation, global warming and its effects on ecosystems, loss of biodiversity, threatened and endangered species, and on and on. Some of these claims of calamity are incontrovertible. A few airplane or satellite pictures provide proof that forests are disappearing throughout much of the world. Other claims are much harder to judge, however. We can't see spotted owls or endangered insects from a satellite, and these pictures don't tell us much about biodiversity either. So how do we know whether a species is declining? How do we know how much biodiversity is in an area? In these cases, we must somehow take samples of what we're interested in, and use statistics to draw conclusions from those samples.

To address large-scale and long-term problems such as endangered species or biodiversity, we need to sample over large areas and for a long time. This effort costs quite a bit of money, so we usually can't take as many samples as we'd like to. Therefore, we need to plan a strategy of sampling that will give us the best possible chance for finding out what we want to know, given the limits on how many samples we can take. Among the things we think about when designing a sampling strategy are:

- How many different years should we sample?

- How many years should we leave between samplings?

- How many different locations should we sample?

- What should the distance be between sampling locations?

- Should sampling locations be random or not?

Because we have a limited ability to sample, the above questions are all interrelated. For instance, as you increase the number of years in which you sample, you must sample fewer locations for any given year.

One of the best monitoring programs in North America is something called the *breeding bird survey*. Started in 1966 by the U.S. Fish and Wildlife Service, the survey has grown to cover all of the United States and parts of Canada. Every June, amateur bird watchers get in their cars and drive along a set of selected routes in each state. Each route is exactly 25 miles long, and starts and ends at the same place each year. Along the route, the bird-watchers stop their cars every 0.5 miles, get out, and listen and look for birds for exactly 3 minutes. At the end of the route, they have stopped 50 times, each time for 3 minutes. The bird-watchers compile a list of how many birds of each species they saw along the route. This list is then put together with the lists from all the other routes into a map of bird species abundances across the continent for that year.

This survey has many limitations. Not every route is visited every year. Some areas of the continent have more roads and more bird-watchers than others, so are sampled better. You can surely think of other problems as well. Nevertheless, the large number of routes (over 3000) and the fact that the whole sampling procedure is so standardized makes the breeding bird survey the best large-scale data set on any group of animals on this continent.

In this lab, you will be put in charge of a small part of the breeding bird survey. You will be asked to look especially hard for two species of birds that ornithologists are afraid might be declining in numbers, and perhaps headed for extinction. In order to do this, you will be given some funds to add more routes to the survey, to be used specifically to look for these two species. You will have to design a sampling scheme with these extra routes, and decide how you are going to analyze the data you collect, in order to determine whether or not these two species of birds are declining in abundance.

Outline of This Lab

You have just been put in charge of the Northwest U.S. section of the breeding bird survey. Along with coordinating the general survey, you have been asked to pay particular attention to two species of woodpeckers, the Three-Toed woodpecker and the Pileated woodpecker. Three-toed woodpeckers are black and white woodpeckers with some yellow in their crown, and through evolution they have lost their hind toes (thus the name). Pileated woodpeckers are large woodpeckers with dark bills and pointy red crests. Both of these birds live for several years, and don't migrate, but they do have large home ranges, which they maintain from year to year. Some ornithologists have suggested that the abundance of these two woodpecker species is declining, particularly in areas that are being converted to farmland or developed in some other way. The Departments of Wildlife of several states in the Northwest are concerned about this decline, as are local environmental groups, and both wish to know whether the decline is real. Because of this concern, the state of Washington has given you some extra money to add routes into your survey to look specifically for these woodpeckers.

The ornithologists who noticed the decline have suggested that you concentrate your sampling effort on one particular part of Washington where both species are relatively abundant. Though this area is now wild, most of it is slated for development over the next 12 years. The money you have been given is enough to run 50 extra routes in this area. Both the government and the environmentalists are willing to give you ten years to figure out whether the populations are declining. This means that you can put in an extra 50 routes spread out any way you want to over the next ten years. Your goal is to decide where and when to take each of these samples, then analyze your results and determine whether the Three-toed or Pileated woodpeckers are indeed declining in abundance as the land becomes developed. This sampling and analysis must be thorough enough to convince the state governments, the ornithologists, and the environmentalists that your conclusions are correct.

I have designed a simulation of these two birds in the area where the ornithologists suggested you sample. Just as in real life, you will only be able to get information on the birds through sampling; you won't be able to see anything on the screen. Although in real life you would have some idea of where development is taking place, I am not showing you that either, to simulate your lack of knowledge of all the other variables that might also be important, but which you cannot see. You will be able to place samples anywhere you want to within the area, and you'll be able to space your samples out any way you want to over the ten years, the only restriction being that you can take a maximum of 50 samples. Each sample tells you which birds are found in a long skinny rectangle, simulating the information you would get by driving along a road. In this simulation, you have perfect bird-watchers going along these routes for you, and they never make mistakes. If they report that there are seven birds along a certain route in a certain year, you can be sure there are seven birds there. The weather also stays constant throughout the ten years.

When you finish, you will write a short briefing on your results to the various interested parties (the state of Washington, environmental groups, and the ornithologists who gave the original warning about population declines). These groups have conflicting interests, so you will have to make your report quite thorough and airtight to convince them all that your results are correct. In addition to talking about your data, you should probably plot it graphically. When you make these plots, note that the "means" of the data are not the only interesting information. It's also interesting to look at how much scatter there is around the mean (the variance, in statistical jargon). The lower the scatter around the mean, the easier it will be to convince people that your results are correct. You can show scatter either by plotting every data point, or by plotting the mean and then putting little error bars around the mean that show plus and minus one standard deviation. To get standard deviation, find the mean for all your samples in a given year, then for each sample, subtract the mean, square the result, add all of these together, divide by the total number of samples minus one, and take the square root. For instance, if you had 5 samples where you found 1, 2, 3, 4, and 5 birds:

The mean is 3.

1–mean = –2	–2 squared = 4
2–mean = –1	–1 squared = 1
3–mean = 0	0 squared = 0
4–mean = 1	1 squared = 1
5–mean = 2	2 squared = 4
Total	10

Total/(number of samples–1) = 10 / 4 = 2.5

Standard deviation = $\sqrt{2.5}$ = 1.6

Other variations you might want to try: plotting the difference in sampled population size from each sampling year to the next; plotting the number of samples in which you found or didn't find a woodpecker of a given species; plotting the maximum and minimum number of woodpeckers you found in a sample each year; or making up your own statistic. Use your creativity, but make sure that you have something you are confident about in the end.

The Lab

1. Run EcoBeaker

> In order to give you practice sampling before you do it for real, I have constructed a practice model for you.

2. Open the situation file Bird Survey Practice (use the OPEN command in the File menu).

> After the situation loads, you should see two windows laid out on the screen as follows:

A map of the area you'll be sampling for birds, which you'll use to place your samples.

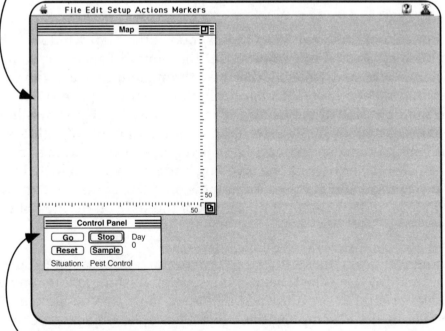

The control panel used to control running and stopping the model

3. Run the model (press the GO button on the Control Panel).

   This model has a bunch of birds of an undetermined species flying around randomly. At any time, you can stop the simulation and take a sample as described below.

4. To take a sample, stop the model (STOP). Then push the SAMPLE button on the Control Panel. Now move your mouse so that you are pointing at the top-left corner of the place you want to sample. You should see a rectangular outline of the area to be sampled following your mouse. You will also see the coordinates that your mouse is pointing to in a small window below the Map. When you reach the spot you want to sample, click the mouse button. A dialog box will appear asking whether you really want to sample that spot. Click on OK. Another dialog box will appear telling you how many birds were found in that sample.

   Repeat step 4 until you are comfortable using the sampling technique. You can also run the model some more to change where the birds are, and you can try different sampling designs on these birds to see what works and what doesn't. In case you are curious, there is a total of 100 birds here.

5. When you are comfortable with the sampling, open the situation file Bird Survey.

   There are only two windows in this model, the Map window, which you will use to place your samples, and the Control Panel. The Map is completely white, and will stay that way throughout this lab. You cannot get any information on what's happening inside it except through sampling. On the right side of the Control Panel is a display that shows how many months have gone by since the model started running. Month 1 is January, 2 is February, and so on, until 12, which is December, and then month 13 is January of the next year. The breeding bird survey is conducted in June, so you will take your samples on months 6, 18, 30, and so on, up to the last sample on month 114 (or some subset of those if you decide not to sample every year).

   Before you go on, make sure you have decided how you are going to sample. Last time I checked, neither the state of Washington nor any environmental group I know of owned a time machine, so you are only going to get one chance at sampling these birds. Once the ten years are up, you are out of time and that's all the data you'll get. Furthermore, the 50-sample limit is strict (only prisons and the federal deficit are allowed to go over budget nowadays). Once you reach 50 samples, you will not be able to take any more. So be careful on your first and only try.

For your information, each sample looks at 10 squares on the Map. The Map is 50 squares wide by 50 squares high, for a total of 2500 squares.

6. Run the model (GO) until month 6. Then STOP the model.

7. Take whatever samples you decided to take in the first year.

The dialog box showing the number of each species looks like this:

Bird   2   Bird   1

The Bird designated on the left is the Pileated woodpecker and the Bird on the right is the Three-toed woodpecker. So the number on the left is the number of Pileated woodpeckers and the number on the right is the number of Three-toed woodpeckers in that sample.

8. Continue running the model (GO), stopping it every 12 months and sampling according to whatever sampling design you came up with.

9. At the end of your sampling, conduct some sort of analysis to determine if either bird species is going extinct.

10. When you are finished, write a short (maximum two pages, not including graphs) briefing on your results. Remember, this briefing must convince several different groups (with differing interests) that your results are correct.

EcoBeaker Manual

Installation

EcoBeaker includes several different installation options. If you are a student or a teacher using the prewritten laboratories included in this manual, then the default installation described here is what you want. If you are a teacher who wants to modify some of the included laboratories, or you want to add to EcoBeaker by programming your own procedures, then you should look at the "Special Installations" sections just below. You may also want to read the "Getting Updates," "More Laboratories," and "Other Files" section, which explains where to get bug fixes, laboratories that are new or didn't make it into this book, and files you will need for programming.

You should run EcoBeaker from a hard disk or a network, not from a floppy diskette. As mentioned on the package, to use EcoBeaker you need, at minimum, a Macintosh or PowerPC computer containing:

- 5 megabytes of memory (8 megabytes if you are running System 7.5 or higher)

- System 7.0 or higher

- 12" or larger color monitor. (EcoBeaker will work on smaller or black and white monitors, but will be quite difficult to use.)

- 2.5 megabytes of free hard disk space, 1.5 megabytes of which will be used for installation; 1 megabyte should be free for temporary files created as EcoBeaker runs.

Default installation of EcoBeaker:

- Insert the EcoBeaker diskette into the disk drive of your computer.

- There is only a single icon on this diskette. Double-click on this icon to run the installation program.

- A window will appear with some text and three buttons at the bottom. Read the text, as it may contain instructions or information that didn't make it into the manual.

- In the lower-left corner is an area labeled Install Location. Select the hard disk (and folder if you wish) where you want the installer to put the files.

- Click on the INSTALL button.

- Wait

EcoBeaker will now be installed on the disk you indicated, ready to use.

I will periodically update EcoBeaker as I fix bugs and add features. See "Getting Updates," "More Laboratories," and "Other Files" below for how to get these updates.

Special Installations

I have included on the disk the full text of the laboratory guide of this manual, in Microsoft Word 5.1 format. I have also included a number of files and a manual for those of you wishing to do your own programming within EcoBeaker. In addition, the default installation tries to be smart about which version of the program it installs (68K or PowerPC), but you may want to install a different version. To get any of these off the installation disk:

- Insert the EcoBeaker diskette into the disk drive of your computer.

- There is only a single icon on this diskette. Double-click on this icon to run the installation program.

- A window will appear with some text and three buttons at the bottom. Read the text, as it may contain instructions or information that didn't make it into the manual.

- Click on the CUSTOM button. This will bring up a new dialog box with a list of all the possible things you can install.

- From this dialog box, select those items you want installed. You may need to do this one at a time. Whenever you select an item, a bit of text at the bottom of the dialog box explains exactly what that item will install. If you want everything on the disk installed, including both versions of the program, the lab guide, and the files for programmers, select FULL INSTALL.

- Click on the OK button. The second dialog box will disappear.

- In the lower-left corner is an area labeled Install Location. Select the hard disk (and folder if you wish) where you want the installer to put the files.

- Click on the INSTALL button.

- Wait

Getting Updates, More Laboratories, and Other Files

I am distributing bug fixes and other updates to the program through my web site. In addition, this web site contains laboratories which didn't make it into this book or are not yet adequately tested, as well as files that you will need for programming in EcoBeaker but wouldn't fit on the disk. The site also contains helpful hints, answers to commonly asked questions, and whatever else I think might be useful for users to have. To get there, point your web browser to:

http://www.webcom.com/sinauer/ecobeaker.html

Tutorial

EcoBeaker is designed to be easy to use, so hopefully it shouldn't take you much time to get up and running with it. This section is a tutorial intended to help you in your initial explorations of EcoBeaker. The tutorial goes through a simple laboratory like those in the lab manual, but it in great detail. If you are one of those people that likes to just play around until you figure a program out, then skip this section and simply use the rest of this manual as a reference when you get stuck. On the other hand, if you want step-by-step instructions to take you very carefully and gingerly through the program, this section is for you. If you are only using one of my prewritten laboratories, you should be able to figure out everything by reading the laboratory instructions, and you probably don't need to look at this tutorial (even if you are a computerphobe). This tutorial and the rest of this manual assumes some familiarity with how a Macintosh works (i.e. it assumes you've run a program on a Mac before), so if you're new to Macintoshes or to computers in general, you might want to find another source to become familiar with the computer before continuing with EcoBeaker.

EcoBeaker is a program to run ecological models. The models in EcoBeaker run in a two-dimensional world, called the Grid because it is split up into little squares. Think of the EcoBeaker Grid as a sort of chess board. Each square can contain a creature (one of the chess pieces), and the creatures in a given model are split up into different types. EcoBeaker calls these types "Species", with the obvious implication that each species is different from other species and that all creatures of a given species are similar to each other. The species are the rooks, knights, pawns, and so on. An EcoBeaker model can have up to ten species in it. One of the species will always be the Empty species, representing squares where there are no creatures. Each species is given a set of rules, which all creatures of that species will obey (for example, pawns can only move forward, rooks can go forward/backward and left/right). You'll see what kinds of rules these can be and how they work below.

First, run the program.

- Find the EcoBeaker program and double-click on its icon to run it.

First to appear will be an introductory screen, which you can get rid of by clicking the mouse button. Then three windows will appear. In the upper left you will see the Species Grid. This is a picture of the modeling world, showing the positions of all the different creatures in the model (it's the chess board). Right now, the only thing in the model is empty space, which in this case is colored black, so the whole grid is black.

In the upper right is the Species Setup window. This is where you will design models, and where you will go to change parameters in a model. On the right side of the Species Setup window is a list of all the species currently in the model. As you can see, there is only one species currently in the model, the Empty species, which represents parts of the grid with nothing in them. Next to the list of species are a number of buttons that let you make new species, delete species you don't want anymore, and modify existing species.

Down towards the bottom of the screen is the Control Panel. This window includes buttons that let you perform common actions in EcoBeaker without going to the menus. You can also get to the same commands from the Action menu. The GO button starts the model running, STOP stops the model without resetting it, RESET stops the model and resets it back to its original state at time 0, and SAMPLE takes a sample with the current sampling technique. To the right of all these buttons is a display showing the current Timestep. When you first start the program or load a new model, you will see that it starts at Timestep 0. As you run the model, the Timestep will go up to 1, then 2, then 3, and so on. Each time the Timestep goes up by one, EcoBeaker activates all the creatures in the model and has them do whatever their set of rules tells them to do. This is known as *running in discrete time*, and is explained more fully in the "How EcoBeaker Works " section of this manual.

- Try it—click on the GO button. You'll see the Timestep increase rapidly. Then click on the RESET button to reset the model back to time 0.

Right now, the model has only empty space in it, and the empty space isn't supposed to do anything, so when you run the model nothing happens other than time going by. This is pretty boring. To make things a little more exciting, let's load a simple model. In EcoBeaker, a model and everything associated with it such as graphs and sampling techniques is saved into something called a Situation File. You can load a situation file just like you would load a file in any other program, using the OPEN command.

- Select the OPEN... item from the FILE menu. A standard Macintosh file requester will appear.

- Select the situation file called Random Settlers, and then click the OPEN button. You will see windows appearing and rearranging themselves on the screen for a second or two, and when things settle down, you will notice the same three windows we discussed above, plus a fourth window called Population Graph.

The situation Random Settlers is a model with one species in it, a plant whose seeds get blown onto the EcoBeaker world. Each seed that comes in settles down and grows up in random location. A certain number of seeds are blown onto the grid each timestep, and every seed that's blown in manages to grow up into a plant, as long as it doesn't land on top of an already existing plant.

- Just to get a little action, try running this model (push the GO button in the Control Panel).

You will see the grid start filling up with green squares. Each green square represents one creature of the Plant species that has settled down onto the grid. Look over at the Population Graph window. Here you'll see a graph showing how many plant seeds have settled over time. Notice that it starts at 0, and rises pretty quickly for a while, and then the speed at which the plant is growing starts leveling off.

- Click on the STOP button.

Notice that the timesteps stop increasing, and no more plants are settling in. You have just stopped time—congratulations, quite an accomplishment don't you think? But better than that, you can make time go backwards.

- Click on the RESET button.

Notice that the timestep has now gone back to 0, and all the plants are wiped off of the grid, leaving the whole world empty as it was when you originally loaded the model. The graph is also reset. You could now run the model again, and see if the pattern is the same as it was the first time you ran it.

Let's say you want to see what happens when you change the number of seeds being blown onto the grid each timestep. A good course of action to follow whenever you change something is to first predict what you think will happen, then change it and run the model and see if you predicted correctly. So make a prediction of what you think will happen when you double the number of seeds being blown into the world. Write this prediction down, or draw out what you predict the Population Graph will look like. Now you must get to the dialog box where the plant species are defined to make your changes.

- Find the name Plant in the Species Setup window and double-click on the name (double-click the mouse button).

This will bring up an imposing-looking dialog box, called the Species Setup box, which defines the rules that the Plant species follows in this model. Although there's a lot of stuff here, it's not too complex if we look at it piece by piece. First, notice that there are four main parts defining the rules of this species, marked off by lines running through the dialog box. On top are the name and color of the species. Below that is a settlement procedure, an action procedure, and a transition matrix. For now, we will ignore the transition matrix (it's not doing anything in this model anyway). If you look where it says Action Proc, you'll see that the current action procedure is None, so we can ignore that also. The only rule for the Plant species, then, is the Settlement Proc, which is given as Fixed. You can look up what the Fixed settlement procedure does in the "Settlement Procedures" section of this manual. Basically, all it does is randomly throw a certain number of new creatures of this species onto the grid each timestep. The number of new creatures per timestep is a Parameter of the Fixed settlement procedure. To change a parameter of a settlement procedure, you must click on the SETTLEMENT PARAMS button.

- Click on the SETTLEMENT PARAMS button in the middle left of the Species Setup box.

When you click on this button you will be presented with a second dialog box; this one has just two items in it. The important item for us is the item called Num Immigrants/Turn, which determines the number of settlers coming onto the grid per timestep. Currently there are 10 new settlers per timestep. Increase this to 20 as follows.

- Find the Num Immigrants/Turn item in the dialog box. This is set to 10. Click on the "10," and use the delete key to erase it. Then type in the number "20".

- Click on the OK button. This will return you to the full Species Setup box.

- Click on OK in the Species Setup box as well. Now you are back to the model.

Before you do anything else, write down your prediction about what will happen. (I know at least half of you ignored this step the first time). What will happen? Will the maximum population size increase? Will the speed with which that maximum is reached increase? Will the shape of the curve in the graph change?

- Now click on the RESET button to reset the model to time 0.

- Run the model (GO).

What happened? Was this what you predicted? If you aren't comfortable with changing parameters yet, try the whole thing again, this time setting the number of settlers to 5 .

Suppose we want to see what will happen with a different settlement procedure. For example, right now all the seeds are coming from outside EcoBeaker's world at a fixed rate. Suppose instead we want a few seeds to come from outside, but most of the seeds to come from plants within the world, with each plant that's growing in the grid producing a certain number of seeds per timestep. There is another settlement procedure called Density Dependent, whose rules are exactly what we want.

To change to the Density Dependent settlement procedure:

- Double-click on the Plant species in the Species Setup panel as before.

The species dialog box for plants will appear. Notice that the current settlement procedure, Fixed, is surrounded by a box that has a drop shadow. This drop shadow indicates that there is a pop-up menu here from which you can change the settlement technique.

- Click within the box surrounding Fixed settlement and hold down the mouse button. A menu will pop up.

- Still holding down the mouse button, move the mouse to highlight DEN-SITY DEPENDENT, and then let go of the mouse button.

You will see that the settlement procedure is now Density Dependent.

- Now click the SETTLEMENT PARAMS button to set the parameters of the new settlement procedure.

This procedure has three parameters. The first is the number of settlers coming in from the outside, which is exactly the same as the Fixed settlement procedure. For now, set this to 1, so that one new seed will be blown in from the outside every time step. The second parameter determines how many seeds will be produced from each plant that is already on the grid. Set this to 0.1, meaning that each time step, one out of ten plants on the grid will produce a seed. These seeds will then land randomly around the grid, exactly like the seeds coming in from the outside. Again, look up the Density Dependent settlement procedure in the Settlement Procedures section of this manual for more information.

- Set Num Immigrants/Turn to 1, and Num Settlers/Individual to 0.1.

- Click on the OK buttons in both dialog boxes.

Now you're back to familiar territory. Remember to make a prediction of what you think will happen. Will the maximum number of plants change? Will the shape of the population curve change? Will the plants reach their maximum number sooner or later? Reset the model, run it (GO), and see what happens.

Finally, let's say you want to add in an herbivore that eats the plant. To do that, you need to make a new species.

- Find the Species Setup window on the right side of the screen, and click on the NEW SPECIES button.

This will bring up a Species Setup box for a newly created species called Species 2. You must do several things to set up the new species. You must give it a name and a color. Since you want it to be a mobile species, you must make it Individualistic (a somewhat subtle thing which is explained in the Grid-Based vs. Individualistic section of the manual). Then you must select a settlement and action procedure for the herbivore.

- Click in the text field labeled Name, delete the default name Species 2, and type in the name you want for the herbivore species.

- Find the colored box in the upper right of the dialog box labeled Color. Click once in this box. A standard Macintosh color wheel will appear and you can select the color you want for this species.

- Click on the INDIVIDUALISTIC check-box so that it's checked.

To make an herbivore, I suggest using the Fixed settlement procedure described above (to get new herbivores into the modeling world), and the Predator action procedure. The Predator procedure will make each herbivore look around for food each timestep, move towards the closest food it sees, and eat it. Each time it eats it will gain some energy, and when it acquires enough energy it will have a baby. As the herbivore moves around, though, it will lose energy, and if it loses all its energy, it will die. See the Individualistic Action Procedures section of this manual for a further description.

- Click and hold down the mouse button in the pop-up menu for selecting the settlement procedure. From this menu, select the FIXED settlement procedure.

- Click on the SETTLEMENT PARAMS button, and set the Num Immigrants/Timestep to 0.1, so that, on average, one new herbivore will enter the grid every ten timesteps.

- Click and hold down the mouse button in the pop-up menu for selecting the action procedure at the bottom of the dialog box, underneath Action Proc. Currently, the action procedure should be None.

- Select the PREDATOR action procedure.

- Click on the ACTION PARAMS button to bring up a dialog box where you can select the parameters for the Predator procedure.

I don't want to repeat myself too much, so for a full description of the all the parameters mentioned here, please see the "Predator in the Individualistic Action Procedures" section of this manual. The default values will give a reasonable behavior to the herbivores, but go ahead and change whatever you want. Then leave the dialog box.

- Click on the OK buttons in both dialog boxes.

As a last step before you run the model, let's add the population size of the herbivore species into the graph so we can compare it to the plant.

- Find the Population Graph window and double-click anywhere within it. Another kind of complicated dialog box will appear. This one is explained in the Graphs section of this manual.

- Click on the ADD ITEM button. This will bring up a second dialog box where you can select the statistic you want to add to the graph.

- At the top of the second dialog box is a pop-up menu where you can select between the different types of statistics in EcoBeaker. From this menu, select SPECIES POPULATION.

When you select the type of statistic you want, the parameters for that statistic will appear in the bottom of the dialog box. For the Species Population statistic, the only parameter is the species whose population size you want. This is selected using a pop-up menu.

- Find the pop-up menu labeled Species and select your herbivore species.

- Click on the OK buttons in both dialog boxes.

The graph will now plot the population size of the herbivore species along with that of the plant species.

- Run the model (GO).

That's the basics of building and running models in EcoBeaker. To follow are more detailed instructions on each of the commands, including how to build your own models from the Species Setup window, how to make your own graphs and change the appearance of the graphs, how to make habitats for the creatures, how to sample the creatures in the model and compare these samples to the real population sizes and distributions, and more. However, most of the things you can do will follow the same basic format. There will be a separate window for the graphs, which will look like the Species Setup window, and another window for Habitats. From within these panels you will have to change parameters and make selections from pop-up menus, exactly as above. These procedures are all explained below, but you can just start trying out the labs and you should quickly get the hang of it. Then you can use this manual mainly as a reference.

I hope that you enjoy playing with EcoBeaker, and I welcome any feedback that you can give me. Have Fun!!!

Action Commands

There are four commands that you will commonly use to make things happen in EcoBeaker. These are the commands to START, STOP, and RESET models, and to SAMPLE. All four commands are included as buttons on the Control Panel, a window that will always be visible while running EcoBeaker. You can also find these commands in the Action menu if you have some aversion to buttons.

You can figure out whether a model is started or stopped by looking at which button is highlighted on the Control Panel. Below is a picture of the Control Panel:

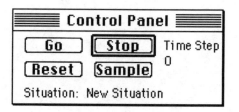

Go

This command starts a simulation running. It will also continue a simulation that is stopped.

Stop

Like Superman, you can stop the world from turning with just one finger. When you select STOP, any currently running simulation will be stopped. You can continue the simulation by selecting GO again. The STOP button will be highlighted while the simulation is stopped, and the timesteps won't advance.

You can do many things within EcoBeaker without stopping the current simulation. You can change parameters, make graphs, even add species, while the simulation is running. Be aware, however, that some changes made to a model while it's running may reset parts of it back to time 0—this is especially true when changes are made to individualistic species.

Reset

This command will reset everything back to the Big Bang (time 0). Everything that was stored about the current run of the model will be erased, and any other initialization needed will be done. Be careful.

Sample

Sampling is what ecology is all about, and EcoBeaker lets you participate in a whole array of ways to sample the Grid before you. (Never mind that you know the actual populations already. After all, what are models for if not to find out things you already know?) Setting up the sampling technique is discussed in the "Sampling" section of this manual. Once a technique is set, use the SAMPLE command to take a sample for you.

Loading, Saving, and Copying

EcoBeaker allows you to extract and import data files in several formats. Perhaps the most common files you'll want to load are situation files. A situation file includes an EcoBeaker model, associated graphs, the current sampling technique, the positioning of windows on the screen, and everything else that can be set in EcoBeaker. When you build a new model, you will save it as a situation file, which can later be loaded back into EcoBeaker.

In addition to situation files, you can save the current state of the species or habitat grids while running a model, as either a graphic or a text file. You can also save data from graphs, or even the graph itself.

Opening a Situation

If you're not feeling creative enough to make your own EcoBeaker world today, you can easily borrow someone else's.

To open a situation file, select the OPEN... command in the File menu. When you select this command, you will be presented with a standard Macintosh file requester, where you should select the situation that you want to load. It will usually take a couple of seconds; you'll see windows bouncing around, then the situation will be ready to roll.

Saving a Situation

As mentioned above, EcoBeaker saves models and all other setup info into situation files, which you can load using the OPEN command described above. To save a situation:

- Set up your model, graphs, and all other settings exactly as you want them.
- Select the SAVE SITUATION... command from the File menu.
- In the dialog box that appears, type in a name for the situation file.
- Click on the SAVE button.

This situation will now be saved in the file you specified. When you load this file back in, the screen will look exactly as it did when you saved it, and everything about the model will be set exactly as before.

If you are not at time 0 when you save a situation, you will be given the option to save not only the model, but also all the creatures, habitats, data in the graphs, and so on, exactly as they are at the time you are saving. If you say YES, all data in the model necessary to exactly reconstruct that time point will be saved in the file. This can be useful, for instance, when it takes a while for a model to stabilize, and you want to save it after it has stabilized.

If you have Markers set (see the "Markers" section of this manual) when you save a situation, you will be given the option to save the markers along with the situation. This means that when you load the situation back in, the markers will still be there.

Saving data and saving markers may significantly increase the amount of disk space that a situation files takes.

Saving Data

EcoBeaker gives you a range of tools for analyzing and graphing data generated by its models. However, it is by no means a full-fledged statistics or graphing program, so you may want to export data from EcoBeaker to another program. EcoBeaker can save the data in the Species Grid, the Habitat Grid, or any of the graphs. The files saved will be text files, which can then be read into almost any other data analysis program. You can also copy and paste data from EcoBeaker into another program (see "Copying Data" below).

To save data from EcoBeaker:

- Select the window whose data you want to save. For instance, if you want to save the Species Grid, click in the species grid window so that it is the frontmost window (its title bar should not be ghosted).
- Select the SAVE DATA... command from the File menu.
- In the dialog box that appears, type in a name for the data file.
- Click the OK button.

The data will now be saved into that file.

If the file you are saving to already exists, you'll be asked whether to overwrite the old file or append to it. If you append, then you can save statistics from several runs of a model in a single file, which you may find convenient.

The saved file is pure text, so you can open it from any other program that can read text files, such as a spreadsheet or graphing program.

The format of the data file will vary depending on what you are saving. If you are saving one of the grids, the data file will contain a matrix of numbers in the same order as the data on the screen. Each row of numbers corresponds to the same row of squares in the grid on the screen.

If you are saving a line graph or a table, the file will contain a line at the top giving the name of the graph, then a second line giving a title for each column of data, and then the data. The first column will be the time for each data point. Subsequent columns will contain each series of data in the order that the data appear in the graph. The whole file will look like a table looks on screen.

For other types of graphs, the file format will be the same as for tables and line graphs, except there will only be one row of data.

For more information on data files, look in the sections on the grids and on graphs.

Copying Data to the Clipboard

As an alternative to saving data from EcoBeaker into a file (as described in "Saving Data" above), it may be more convenient to copy data from a part of EcoBeaker into the clipboard, and then paste this data into another program. EcoBeaker will let you copy the data contained in the Species or Habitat Grids, or in any of the graphs.

To copy data from one of these places into the clipboard, first click somewhere in its window to make the window active. Then select the COPY DATA command in the Edit menu. The data shown in the currently selected window will now be stored as text in the clipboard. Look in the "Saving Data" section above for a description of how the data is formatted. You should be able to paste this data into any spreadsheet or graphing program.

Saving Pictures

You can save pictures of the Species Grid, the Habitat Grid, or any graph into a file, which can then be read into almost any painting or drawing program (it's a PICT file). The saved pictures will look exactly like what you see on the screen.

To save a picture of a grid or graph:

- Select the window that contains the picture you want to save by clicking in that window (so its title bar is not ghosted).

- Select the SAVE PICTURE command from the File menu.

- Type in a file name for the picture file.

- Click the OK button.

The picture will now be saved in this file, which you can then load into any program that can read PICT files.

Copying Pictures to the Clipboard

Instead of saving pictures of the grids or the graphs as described above, it may be more convenient for you to copy them to the clipboard, run a drawing program, and paste the pictures into the drawing program. You can copy a picture of the Species or Habitat Grid, or of any graph. To do this, select the window that contains the picture you want to save by clicking in that window (so its title bar is not ghosted). Then select the COPY command from the Edit menu. The picture will be put into the clipboard, and you can now switch to another program and PASTE the picture into that program.

Printing

I've had a little trouble figuring out how to get EcoBeaker to make useful print-outs, since so much of EcoBeaker is based on using color, and most printers only print in black and white. Still, since printing is quite useful, I've put in a simple routine that takes what you see on the screen, sends it to the printer, and hopes for the best. Using this printing routine, you can print a picture of the Species Grid, the Habitat Grid, or any graph.

Printing a Grid or a Graph

To print one of the grids or a graph:

- Select the grid or graph window by clicking in that window (so its title bar is not ghosted).

- Select the PRINT... command from the File menu.

- In the Print dialog box that appears, click on the OPTIONS button. In the Print Options dialog box, set the printer to Color/Grayscale. This setting may not be available for all printers, but if it is, it makes the output look much better.

- Click OK in all the dialog boxes.

The contents of the window you selected will now be sent to your printer exactly as they appear on the screen. The printed output will be the same size as the window on the screen, and the print will be located in the upper left-hand corner of the page.

Page Setup

You can use the PAGE SETUP command to do the standard printing manipulations that Page Setup does, such as controlling various ways the printer processes the pictures sent to it, shrinking and expanding printed pictures, and turning the page sideways.

Constructing Models

One of the nice things about EcoBeaker is that you can quickly construct your own models. While these models may need additional work to be research quality (see my "Note to Researchers"), I manage to make the rough cut on quite a wide variety of models using only the procedures built into the program. The next few sections of this manual explain the components you will need in order to make an EcoBeaker model.

Constructing models is done primarily by making species and assigning each species a set of rules that governs its behavior (the settlement and action procedures). You may also want to add in habitats to the model. In addition, you may change characteristics of the grid(s). Those are the three basic components of all models in EcoBeaker. Once these are set, you can add in graphs to see what is going on as the model runs, and a parameter window to make it easier to change key parameters. As final polish, you can also change the names of some objects in EcoBeaker to names suitable for the model you are constructing. Once you set up all the individual pieces, the model will be constructed and you can run it using the action commands discussed above.

Constructing EcoBeaker models is usually an iterative process. I add in the species I want and set things up a little, then run the partially constructed model. Running it gives me new ideas and points out flaws in my original design, so I go back and change things, maybe adding in or taking out species, changing the size of the grid, changing parameters, and so on. Finally, when I have everything just about how I want it, I do some polishing by arranging the windows on the screen and relabeling different windows.

Here's an example of how I might go about setting up a new model. You certainly do not have to follow this order to get a good model, but perhaps it will help you get started. All these steps are explained in the next few sections of the manual.

1. Select NEW SITUATION from the File menu.

2. Find the Species Setup window, and click the ADD SPECIES button until you have all the species you want in the model. As you're adding in species, give them names and colors but nothing else.

3. If you want to add habitats into the model, select the HABITAT... item in the Setup menu.

4. Find the Habitat Setup window and click the ADD HABITAT button until you have all the habitats you want, again giving them only names and colors.

5. Go back to each species and set its procedures and transition matrix.

6. Go back to each habitat and set its habitat procedure.

7. Change the size of the grid to whatever size you want. Also, perhaps combine species and habitat grid if you have both.

8. Try running the model.

9. Repeat 5 through 8 until the model is running the way you want it to. Also add or delete species or habitats if you need to.

10. Make a Parameter window if you want one.

11. Move all the windows around on the screen until they look good. Also change window titles so that they are appropriate.

12. Fine-tune colors of all species and habitats so that they look good.

13. Distribute to lots of hapless students to find the rest of the bugs (just kidding, I'm sure...).

If you plan to use a model for some serious purpose such as research or consulting work, you should read my "Note to Researchers" at the beginning of this manual. You will then probably want to program your own procedures so that they are exactly what you want. For most teaching purposes, however, the program as is should be quite flexible.

Species

The species is the basic unit in EcoBeaker models. The word species has a lot of connotations in biology, but EcoBeaker uses the term very loosely. An EcoBeaker species is simply a collection of rules governing the behavior of members of that species. Thus a species in EcoBeaker could be an age class, a sex, or even something as unbiological as fire.

There can be up to ten species in a model, and the model can have any number of individuals (or creatures) from each species alive at any point in time. One of the species will always be the Empty species, representing squares with no creatures in them. The rules associated with these species make up the basis of EcoBeaker models.

There are three sets of rules that govern each species, the Settlement Procedure, the Action Procedure, and the Transition Matrix. For the first two, EcoBeaker contains lists of procedures from which you will pick one for your species. Once you pick the appropriate settlement and action procedures, you can fine-tune them by setting parameter values. The Transition Matrix is the same rule for each species, so you simply need to set its values.

The first set of rules is called the Settlement Procedure. This says how new individuals of this species get onto the Grid from outside. The Grid is the area in which the model takes place. For most species you can think of this as immigration. For instance, a simple settlement procedure might say that five new individuals of this species are added in random positions every timestep.

The second set of rules is called the Action Procedure. This says what action each individual will take once it's on the Grid. For most species, you can think of this as the creature's behavior. For example, one procedure makes each creature look around for food, move towards the closest food it sees, and eat it.

The Transition Matrix says what the chance is of an individual of this species becoming an individual of another species. You might use this, for instance, in a stage-structured model where there is a certain probability per timestep of progressing from one stage to the next. You can also use this to do density-independent death by setting a probability of a creature turning into an Empty square.

You can mix and match these three sets of rules, as well as set parameters for each one, to form many unique species. The rules specified by each Settlement and Action procedures included in EcoBeaker are discussed later in this manual. The mechanics of selecting settlement procedures, action procedures, the transition matrix, and more mundane aspects of a creature such as its name and color are all discussed here.

Making a New Species

To add a new species to a model, find the Species Setup window. This window is usually titled Species, and initially appears just to the right of the Species Grid window when you first start up the program. If this window is not visible, you can make it appear by selecting the SPECIES... item from the Setup menu. If the window is shrunk down, you will need to expand it. See "Shrinking, Hiding, and Showing the Habitat Setup Window."

Now click on the ADD SPECIES button in the left side of the Species Setup window. This will make a new species, and bring up the Species Setup box for this new species.

The Species Setup box is the most complicated dialog box in the program, but it's arranged in four smaller pieces, each of which is simpler. Each piece is described in more detail below. The top part lets you set the name and color of the species and choose whether the species is Grid-Based or Individualistic (see below). The middle left lets you set the settlement procedure for the species. The bottom left is where you set the action procedure for the species. Finally, on the right side is the transition matrix. The settlement procedure, action procedure, and transition matrix together make up the rules governing this species.

Here is a picture of the Species Setup box:

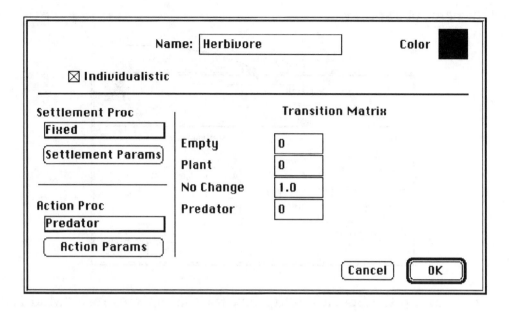

Modifying a Species

To modify a species, find its name in the Species Setup window. Double-click on the species name to bring up a Species Setup box for that species. Then follow the instructions in the next several sections of this manual to change various parts of the species.

If the Species Setup window is hidden, you can always bring it to the front of the screen by selecting the SPECIES... item from the Setup menu.

Naming a Species

To set the name of a species:

- Bring up the Species Setup box for that species (see "Modifying a Species" above).

- At the very top of this setup box is a text field labeled Name. Type in the name you want to give the species here.

- Click on the OK button.

The name of a species can be up to 50 characters long. Names longer than 10 or 20 characters, however, will be unwieldy and get cut off when they are displayed.

Setting a Species Color

To set the color of a species:

- Bring up the Species Setup box for that species (see "Modifying a Species" above).

- In the upper-right corner of this setup box is a colored square labeled Color. Click once on this square.

- A standard Macintosh color requester will appear. Select the color you want for this species.

- Click on the OK buttons in both dialog boxes.

If you plan on having other people use your model, it's worth spending some time picking colors for the different species and habitats that have high contrast with each other. Also remember that many people are color blind, so you may want to avoid certain colors.

Grid-Based vs. Individualistic Species

There are two fundamentally different types of species which EcoBeaker can use. These are called Grid-Based and Individualistic. The difference comes in how each individual creature within that species is stored.

For Grid-Based species, each individual is stored as a color on the Species Grid, and nowhere else. This means that if something else lands on top of that individual and makes the square a different color, the individual is killed. It also means that all individuals of that species are exactly identical to each other, since no information is stored about each individual. Finally, it means that there can be at most one creature from a Grid-Based species on each grid square.

Individualistic species are different in that, in addition to each individual being stored as a color on the Species Grid, there is also a separate list of all the individuals of that species. This separate list contains the position of each individual. It can also contain other information about each individual, so that individuals can be different from one another based on their current state. For example, each individual might have its own energy level, or might store the direction in which it was moving. These additional pieces of information about each individual are called Individual Parameters. Because the positions of individuals are stored separately from the Grid, two Individualistic creatures can also occupy the same grid square.

This distinction between Individualistic and Grid-Based species is further explained in the "How EcoBeaker Works" section of this manual. Depending on the set of rules you want the species to have, it may be advantageous to use one or the other type of species. If you are using only the settlement procedure of the species, then you most likely want a Grid-Based species. If you are using only the transition matrix, or the settlement procedure and the transition matrix, then you also probably want a Grid-Based species, although for certain models using an Individualistic species with the transition matrix might be faster. If you are using an action procedure, you almost certainly want an Individualistic species. The available action procedures are different depending on what type of species you are using, and there is a much larger set of action procedures for Individualistic species.

Note that there is a limit on the number of individual creatures of a given Individualistic species (currently ~4000, though you can increase this to millions—see Increasing the Maximum Number of Creatures), and no limit on the number of individuals of a Grid-Based species (except the number of grid squares).

By default, species are Grid-Based. To make a species Individualistic:

- Bring up the Species Setup box for that species (see "Modifying a Species" above).

- In the upper-left of this setup box is a check box labeled Individualistic. Check this box to get an Individualistic species, and uncheck it to get a Grid-Based species.

- After changing the type of a species, the Action procedure will be reset to None. If you want the species to have an Action procedure, select that (as explained below) and set its parameters.

- Click on the OK button.

Selecting a Settlement Procedure

The Settlement Procedure of a species determines how new individuals of that species enter the model from "outside" the model world. Usually you can think of this procedure as the rules governing immigration for the species. New creatures of a species can also be made "inside" the model world, for instance, when an already-existing creature gives birth to a new creature. The creation of new creatures from inside the model world is governed by the Action procedure, not by the Settlement procedure. Most species will have new creatures coming from both Settlement and Action procedures. You might use the Settlement procedure to put a few initial creatures in the model at time 0, for instance, and from then on have all new creatures come from reproduction governed by the Action procedure.

By default, the Settlement procedure of a new species is None, which does nothing. Descriptions of all the other Settlement procedures are given in the "Settlement Procedures" section in the "Included Procedures" part of this manual. To change the Settlement procedure:

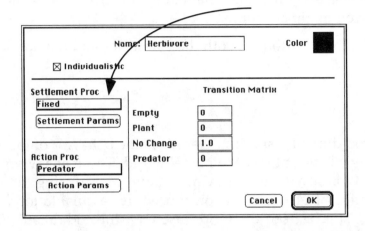

• Bring up the Species Setup box for that species (see "Modifying a Species" above).

• Find the area of the setup box labeled "Settlement Procedure" (in the middle of the left half of the setup box, but it occasionally may be called something different). Click on the pop-up menu and hold the mouse button down.

• A menu will pop up giving all the Settlement procedures currently included in EcoBeaker. Move the mouse to select the one you want, and then let go of the mouse button.

You will now see the name of the procedure that you selected displayed in the pop-up menu. Each procedure has parameters associated with it, and these are all set to default values when you first select the procedure. The default values most likely be incorrect for your model, so you'll probably want to change them. In most cases, you will need to play with some of the parameters for a while before you get exactly the behavior you want. The next section explains how to change the parameters of a Settlement Procedure.

Changing the Parameters of a Settlement Procedure

To change the parameters of a Settlement procedure:

• Bring up the Species Setup box for that species (see "Modifying a Species" above).

• Find the button labeled Settlement Params (in the middle of the left half of the setup box). Click on this button.

• A new dialog box will appear containing all the parameters for this Settlement procedure. Change their values as you wish. See the "Types of Parameters" section of this manual for descriptions of the types of parameters you might encounter.

• Click the OK buttons in both the parameter and the Species Setup dialog boxes.

Action Procedure

The Action Procedure of a species determines the behavior of individuals of that species once they have entered the model world. These behaviors can be almost anything, but typically include moving, eating, reproducing, dying, and so on. There are a different group of Action procedures available for Grid-Based and Individualistic species (see the Grid-Based vs. Individualistic Species section above). If you are using an Action procedure for a species, you will almost always want the species to be Individualistic.

By default, the Action Procedure of a new species is None, which does nothing. Descriptions of all the other Action procedures are given in the Grid-Base Action Procedures and Individualistic Action Procedures sections in the "Included Procedures" part of this manual. To change the Action procedure:

• Bring up the Species Setup box for that species (see "Modifying a Species" above).

- Decide whether you want a Grid-Based or Individualistic species and set the species accordingly (see the Grid-Based vs. Individualistic Species section of this manual).

- Find the area of the setup box labeled Action Procedure (in the bottom left corner of the setup box, but occasionally it may be called something different). Click on the pop-up menu and hold the mouse button down.

- A menu will pop up giving all the Action procedures currently included in EcoBeaker. Move the mouse to select the one you want, and then let go of the mouse button.

You will now see the name of the procedure that you selected displayed in the pop-up menu. Each procedure has parameters associated with it, and these are all set to default values when you first select the procedure. The default values will most likely be incorrect for your model, so you'll probably want to change them. In most cases, you will need to play with some of the parameters for a while before you get exactly the behavior you want. The next section explains how to change the parameters of an Action procedure.

Changing the Parameters of an Action Procedure

To change the parameters of an Action procedure:

- Bring up the Species Setup box for that species (see "Modifying a Species" above).

- Find the button labeled Action Params (at the lower-left of the setup box). Click on this button.

- A new dialog box will appear containing all the parameters for this Action procedure. Change their values as you wish. See the Types of Parameters section of this manual for descriptions of the types of parameters you might encounter.

- Click the OK buttons in both the parameter and the Species Setup dialog boxes.

Transition Matrix

The Transition Matrix specifies the chance per time step of a creature of this species turning into a creature of another species. It can be used in many places where individuals of different "species" turn into one another, such as in age-structured models, in models of succession, and for giving a chance of death (a creature turning into an Empty square). The transition matrix of a species is a list of probabilities, one for each species in the model, including itself. The whole set of probabilities must add up to one.

If you are having a bit of difficulty imagining what the transition matrix is doing, here's another explanation. Let's say you have a model with two species, Grass and Trees, and you are modeling the succession from empty ground to grasslands to forests using only the transition matrix. Each square in the Grid either contains grass, trees, or is empty. During each timestep, EcoBeaker goes through each square in the Grid and looks at what is in that square. If the square is empty, EcoBeaker then looks at the transition matrix of the Empty species. Let's say the transition matrix of Empty is 0.2 for No Change (transition to Empty means that no change took place), 0.6 for Grass, and 0.2 for Trees. EcoBeaker will pick a random number between 0 and 1. If the number is below 0.2, then the square will remain empty. If it's between 0.2 and 0.8 then the square will become Grass, and if the number is higher than 0.8 then the square will become Trees.

Similarly, the Grass species might have a transition matrix specifying a 0.05 probability for transition to Empty, a 0.7 chance for No Change (transition to Grass), and a 0.25 chance for transition to Trees. In each square containing grass, EcoBeaker will pick a random number between 0 and 1, and according to the above transition probabilities either kill the grass (make the square Empty), leave the grass as it is, or replace the Grass with Trees.

By default, the transition matrix of a species does nothing (has a 1.0 probability of No Change, and a 0.0 probability of transition to each other species). To change the transition matrix:

- Bring up the Species Setup box for that species (see "Modifying a Species" above).

- The right half of this dialog box contains the Transition Matrix— a list of all the species in the model with a text field next to each one containing the probability. Type in the probabilities you want, making sure they add up to one.

- Click the OK button.

You can have transitions between creatures belonging to Individualistic species to Grid-Based species, and vice-versa. These will work correctly. However, transitioning from an Individualistic creature to a Grid-Based creature can be a little dangerous if, in your model, two individualistic creatures might occupy the same grid square at the same time. In that case, if one of them transitions to a Grid-Based creature, then the second one moves, the movement of the second one out of the square will erase the first one, and EcoBeaker will not notice this. Just a little warning.

Removing a Species

If you want to remove one of the species in the model:

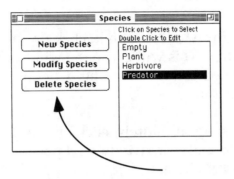

- Find the Species Setup window (the window showing a list of the species in the model). If the window is shrunk, expand it (see "Shrinking, Hiding, and Showing the Species Setup Window").

- Click once on the name of the species that you want to delete so that it's selected.

- Click on the DELETE SPECIES button on the left side of the Species Setup window.

The species will be removed from the model, will disappear from the list of species in the window, and any species of this type in the Species Grid will be turned into the Empty species. Any dependencies that other species had on this species (e.g., if something ate it) will be removed, hopefully intelligently, as will be any graph statistics for this species. You probably want to RESET a model after doing this.

Note that you can't remove the Empty species.

Shrinking, Hiding, and Showing the Species Setup Window

To avoid clutter on the screen, you can make the Species Setup window shrink, or make it disappear altogether.

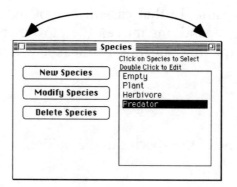

To make the Species Setup window shrink, click on the little icon in the upper-right corner of the window. This will cause the window to shrink in size so that only the names of the species are showing. Clicking on that icon again will expand the window back out.

To make the window disappear entirely, click on the close icon in the upper-left corner of the window. To make the window reappear, you must go to the Setup menu and select the SPECIES... item.

Habitats

In addition to the Species Grid—the area where creatures run around—there is also a second grid within EcoBeaker called the Habitat Grid. As its name implies, this second grid gives you the ability to make models with a varying landscape. EcoBeaker does not include a GIS system—each square in the Grid is of a single habitat type. Despite this restriction, however, the habitats can be used fairly generally to simulate different types of landscapes.

The Habitat Grid works very similarly to the Species Grid. You can have up to ten different Habitats, each with its own set of rules governing what happens to habitats of that type over the course of a simulation. However, habitats are simpler in their construction than species. There is only one procedure governing a habitat's behavior, called a Habitat Procedure. The Habitat procedure works in a manner similar to the Settlement procedure for species. The Habitat procedure of each type of habitat is called once at the beginning of every timestep. In fact, the habitat procedures are the first thing that are done in each timestep.

Setting Up Habitats

To set up habitats, select the HABITAT... item from the Setup menu.

An alert will appear, asking if you want to add habitats to this model. Click on YES.

Two new windows will appear. The first is the window showing the Habitat Grid, initially labeled Habitat Grid. The size of the Habitat Grid window may be different than the size of the Species Grid window, but the dimensions of the Habitat Grid (the number of squares in the grid) always stay the same as the dimensions of the Species Grid.

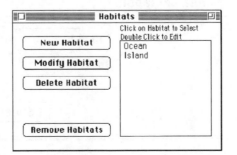

The other new window is the Habitat Setup window, initially labeled Habitats. This is where you will add, delete, and modify habitats.

Making a New Habitat

When you first introduce habitats into a model, the only existing habitat is the Empty habitat, analogous to the Empty species. You will see Empty listed in the Habitat Setup Window.

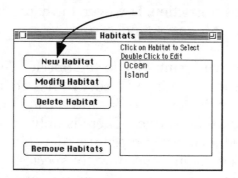

To add another habitat, click on the NEW HABITAT button. This brings up the Habitat Setup box. The Habitat Setup box is the main dialog box for dealing with a habitat. You use it to set the name and color of a habitat, the procedure that a habitat uses, and the parameters for that procedure. All of these are described in the next few sections of this manual.

A picture of a typical Habitat Setup Box is shown below:

Habitat Name:	Island		Color	■

Habitat Procedure: Single Rectangle

Left	1	Top	1
Right	50	Bottom	50

Cancel   OK

Modifying a Habitat

To modify a habitat, find its name in the Habitat Setup window. Double-click on the habitat's name to bring up a Habitat Setup box for the habitat. Then follow the instructions in the next several sections of this manual to change various parts of the habitat.

If the Habitat Setup window is hidden, you can always bring it to the front of the screen by selecting the HABITAT... item from the Setup menu.

Naming a Habitat

To set the name of a habitat:

- Bring up the Habitat Setup box for that species (see "Modifying a Habitat" above).

- At the very top of this setup box is a text field labeled Name. Type in the name you want to give the habitat.

- Click on the OK button.

The name of a habitat can be up to 50 characters long. Names longer than 10 or 20 characters, however, will get cut off when they are displayed and will be unwieldy.

Setting a Habitat's Color

To set the color of a habitat:

- Bring up the Habitat Setup box for that species (see "Modifying a Habitat" above).

- In the upper-right corner of this setup box is a colored square labeled Color. Click once on this square.

- A standard Macintosh color requester will appear. Select the color you want for this habitat.

- Click on the OK buttons in both dialog boxes.

If you plan on having other people use your model, it's worth spending some time picking colors for the different habitats and species that have high contrast with each other. Also remember that many people are color blind, so you may want to avoid certain colors.

Setting a Habitat Procedure

The Habitat procedure of a habitat determines when and where the habitat will be found on the Habitat Grid. Most Habitat procedures turn a section of the Grid into their habitat at a given timestep (usually timestep 0) and then do nothing. However, some Habitat procedures add more of their type of habitat every timestep, and a few do more complicated things such as changing one habitat type to another under certain conditions.

By default, the Habitat procedure of a new habitat is None, which does nothing. Descriptions of all the other Habitat procedures are given in the "Habitat Procedures" section in the "Included Procedures" part of this manual. To change the Habitat procedure:

- Bring up the Habitat Setup box for that habitat (see "Modifying a Habitat" above).

None		
Habitat Name: I	Change on Cycle	Color
	Periodic Add Rectangle	
Habitat Procedure:	Random Circles	
	Regular Circles	
Left 1	Regular Circles 2	
	Regular Rectangles	
Right 50	Regular Rectangles 2 I 50	
	Single Rectangle	

Cancel   OK

- Find the pop-up menu labeled Habitat Procedur" (in the top of the setup box). Click on this pop-up menu and hold the mouse button down.

- A menu will pop up giving all the Habitat procedures currently included in EcoBeaker. Move the mouse to select the one you want, and then let go of the mouse button.

You will now see the name of the procedure you selected displayed in the pop-up menu. Each procedure has parameters associated with it, and these are all set to default values when you first select the procedure. The default values will most likely be incorrect for your model, so you'll probably want to change them. In most cases, you will need to play with some of the parameters for a while before you get exactly the behavior you want.

To change the parameters of a Habitat procedure, do the following:

- Bring up the Habitat Setup box for that species (see "Modifying a Habitat" above).

- At the bottom of the setup box are the parameters for this Habitat procedure. Change their values as you wish. See the Types of Parameters section of this manual for descriptions of the types of parameters you might encounter.

- Click on the OK button.

Removing a Habitat

If you want to remove one of the habitats in the model, do the following:

- Find the Habitat Setup window (the window showing a list of the habitats in the model). If the window is shrunk, expand it (see "Shrinking, Hiding, and Showing the Habitat Setup Window").

- Click once on the name of the habitat that you want to delete so that it's selected.

- Click on the DELETE HABITAT button on the left side of the Habitat Setup window.

The habitat will be removed from the model, will disappear from the list of habitats in the window, and any habitat of this type in the Habitat Grid will be turned into the Empty habitat. Any dependencies that other habitats or species had on this habitat will be removed, hopefully intelligently, as will any graph statistics for this habitat. You will probably want to RESET a model after doing this.

Note that you can't remove the Empty habitat.

Shrinking, Hiding, and Showing the Habitat Setup Window

To avoid clutter on the screen, you can make the Habitat Setup Window shrink, or make it disappear altogether.

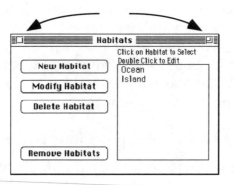

To make the window shrink, click on the little icon in the upper-right corner of the window. This will cause the window to shrink in size so that only the names of the habitats are showing. Clicking on that icon again will expand the window back out.

To make the window disappear entirely, click on the close icon in the upper-left corner of the window. To make the window reappear, you must go to the Setup menu and select the HABITAT... item.

Getting Rid of Habitats from the Model

If you add a Habitat Grid to a model and then decide you don't want it:

• Find the Habitat Setup window (the window showing a list of the habitats in the model). If the window is shrunk, expand it (see"Shrinking, Hiding, and Showing the Habitat Setup" window).

• Click on the REMOVE HABITAT button in the lower-left corner of the Habitat Setup window.

• An alert will appear asking whether you really want to remove the habitats. Click on the YES button.

All habitats will now be removed from the model, and both the Habitat Grid and Habitat Setup Window will disappear. If any species depended on a habitat for something, those dependencies will now be set to the Empty habitat (which still exists even when habitats are not explicitly in the model).

Setting Up the Grids

EcoBeaker models take place in a two-dimensional rectangular world called the Grid. The Grid is composed of an array of squares, and each square generally holds one creature. There are actually two Grids, the Species Grid, which holds all the species, and the Habitat Grid, which holds habitats if a model includes habitats. These two grids are always the same size as each other.

There are several characteristics of the Grids used in EcoBeaker that you can control. Some affect the models, such as the width and height of the grids, and the total number of creatures that can be on a grid at once. Some determine whether the grids are visible on the screen, whether the species and habitat grids are drawn in the same or separate windows, and the rate at which their pictures on the screen are updated. These characteristics are controlled from a dialog box called the Grid Setup box.

The Grid Setup Box

To get to the Grid Setup box, select the GRID... item from the Setup menu.

This will bring up a dialog box that looks like this:

```
┌──────────────────────────────────────────────┐
│                  Grid Setup                    │
│                                                │
│   Grid Width  [50]      Grid Height  [50]      │
│                                                │
│   ● 2-byte Grids    │  ● Non-Flipping          │
│   ○ 4-byte Grids    │  ○ Full-Flipping         │
│                                                │
│   ──────────────────────────────────────       │
│             Appearance of the Grids            │
│                                                │
│   ☐ Hide Species Grid   ☐ Hide Habitat Grid    │
│            ☐ Combine Grids                      │
│                                                │
│   Redraw every  [0]      timesteps             │
│                                                │
│      ( Cancel )          (  OK  )              │
└──────────────────────────────────────────────┘
```

The items in this dialog box are explained in the next several sections of the manual and in the "Controlling the Modeling Algorithm" section. If you don't have habitats in your model the bottom three check-boxes won't appear.

Setting the Grid Size

To set the number of squares across and down the Grid:

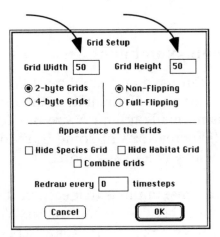

- Select the GRID... item from the Setup menu.

- The Grid Setup box will appear. At the top of this box are two text fields labeled Grid Width and Grid Height. Type the number of squares you want across the grid, and the number of squares from top to bottom, respectively, into these two fields.

- Click on the OK button.

The grid will now be changed to the size you specified. This size is the number of squares in the Grid for the model to use. The total number of squares in the Grid is its width times its height. This size has nothing to do with how big the grid appears on the screen (that's set by the size of the grid windows).

Note that the bigger you make the grid, the more memory it will use up. For small grid sizes (say, 100 by 100 or less), memory size should never be a problem. However, if you want to make a really big grid, you may want to calculate how much memory it will take, and make sure you have that much available.

Each square in the grid takes two bytes of memory (or four bytes if you use a 32-bit grid), so the amount of memory used by the Species Grid will be its width times its height times 2. If there is a Habitat Grid, then you must double this amount, since the Habitat Grid uses the same amount of memory as the Species Grid. You never have a 32-bit Habitat Grid, though.

Typically, EcoBeaker has about 1,500,000 bytes (1.5 megabytes) of memory available. However, other things in a model use memory as well, so it would be wise to leave quite a bit of memory free. See "Memory Matters" below for a further discussion of memory (including how to increase the memory that EcoBeaker has available).

Increasing the Maximum Number of Creatures

Any model in EcoBeaker can have any number of Grid-Based creatures in it at once, limited only by the number of squares in the grid. For an Individualistic species, however, there is normally a limit of around 4000 creatures. This means that if you get above 4000 creatures of some individualistic species in your model, EcoBeaker will not let you add any more.

The technical reason for this maximum is that each grid square is two bytes large. Two bytes can store a number between 0 and 65,536. With a maximum of sixteen species, each species can have 65,536 / 16 = 4096 creatures . (This version can only have ten species, but the next version will have 16.) For most models, this will be enough. I've never reached the limit in a model built for teaching. If you have some huge model, though, you can switch from using a two-byte-per-square grid to a four-byte-per-square grid. This will let you have millions of creatures from each species.

The default grid square size in EcoBeaker is two bytes. To switch the grid from two bytes to four bytes:

- Select the GRID... item from the Setup menu.

- The Grid Setup box will appear. On the left of this box, near the top, are two radio buttons labeled 2-BYTE GRID and 4-BYTE GRID. Click on the '4-byte Grid' radio button to switch to a 4-byte grid, and click on the '2-byte Grid' radio button to switch back to a 2-byte Grid.

- Click on the OK button.

Note that changing back and forth between 2-byte and 4-byte grids will reset your model.

Grid Picture Update Rate

By default, every time something happens as a model is running in EcoBeaker, the change is shown on the screen in the Species Grid or Habitat Grid. This way, you continuously see what's going on in your models. In a complicated model, however, this display can slow down the speed of the model quite a bit, since drawing to the screen is slow. It also can make the model look a bit jerky. So, for example, if your timestep is one year, you might want all the changes from year to year to appear at once.

You can tell EcoBeaker how often to update the picture of the grids on the screen. The default is to update the picture on the screen every 0 timesteps, which means continuously updating the picture. If you change this to updating every 1 timestep, then EcoBeaker updates the screen once at the end of each timestep. If the update rate is 2, the picture of each grid on the screen will only be updated every second timestep, and so on. Changing the update rate from 0 to 1 will significantly speed up some models.

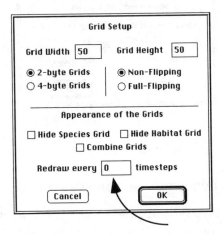

To change the update rate:

- Select the GRID... item from the Setup menu.

- The Grid Setup box will appear. At the bottom of the dialog box is a text field labeled Redraw every ___ timesteps. Fill in the blank with the update rate that you want.

- Click on the OK button.

The update rate has no effect on the model. It only affects how often the screen is updated. Also note that, while you can get a big speed increase from updating the grid less frequently, it probably won't be quite as big as hiding the grid entirely (see below).

Hiding the Species or Habitat Grid

Normally the Species Grid will be displayed as you run a model, as will the Habitat Grid if you have habitats, so that you can see what's going on. Sometimes, however, you may want to hide one of the grids from view. One reason you might do this is to increase speed at which the model is running. Hiding the Species Grid can speed up a model from 20% to 30%, since drawing to the screen is a lot of work for the computer. Another case in which you might hide a grid is to add a little mystery to what's going on. For instance, you might set up a distribution and ask students to guess what it is, through sampling it. A third reason is to avoid cluttering up the screen with too many windows.

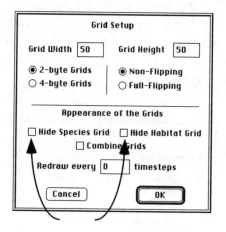

To hide one of the grids:

- Select the GRID... item from the Setup menu.

- The Grid Setup box will appear. Near the bottom of this dialog box are three check boxes (only one if you don't have habitats). To hide the window showing the Species Grid, check the HIDE SPECIES GRID check-box. Similarly, to hide the Habitat Grid check the HIDE HABITAT GRID check-box.

- Click on the OK button.

When you click on OK, the grid(s) you told the program to hide will disappear. Your model will still run normally, and if you want to view a hidden grid again, uncheck the appropriate check-box and the grid will reappear.

Combining the Species and Habitat Grids

If you have habitats in a model, you are usually interested in looking at the position of the creatures in the model relative to the habitats. For some models this relative position is obvious, but for others it's very helpful to be able to overlay the Species Grid on top of the Habitat Grid. EcoBeaker lets you draw both the habitats and the species in the Species Grid window. The habitats will show through in every square that has no creatures in it (every square in the Species Grid that is Empty).To overlay the species and the habitats in the Species Grid window:

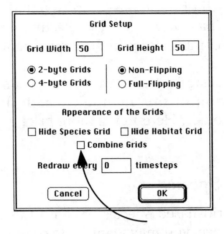

- Select the GRID... item from the Setup menu.

- The Grid Setup box will appear. Near the bottom of this dialog box are three check-boxes. The bottom of these is labeled Combine Grids. Check the COMBINE GRIDS check-box to draw both grids in the Species Grid window. Uncheck it to remove the picture of the habitats from the Species Grid window.

- Click on the OK button.

The Combine Grids setting is convenient not only for comparing distributions of creatures with distributions of habitats, but also for removing the Habitat Grid from the screen (to avoid clutter) without losing the information in it. There are two drawbacks to removing the Habitat Grid. First, some of the habitat will be hidden by creatures so you won't see all the habitats. Second, if you have lots of species and lots of habitats, finding enough colors to distinguish them all will be tricky. Still, when I have habitats in a model, I usually combine the grid pictures, and hide the Habitat Grid window. See the "Hiding the Species or Habitat Grid" section of this manual above for instructions on hiding grid windows.

Controlling the Modeling Algorithm

The algorithm that runs the models is described in detail in the "How EcoBeaker Works" section of this manual. There are a number of details of this algorithm that you can set, depending on the type of model you have made. You can tell it not to process Grid-Based species, which will increase the speed at which a model runs. You can set how much the program randomizes the order in which things act, which is a trade-off between speed and accuracy. You can also set whether the algorithm will flip between two grids on successive timesteps, or do everything on one grid. All of these are set in two dialog boxes, the Grid Setup and Other Setup boxes, which you access by selecting the GRID... (see the "Setting up the Grids" section) or OTHER... items in the Setup menu, respectively.

In order to really understand these options, you should read the "How EcoBeaker Works" section of this manual, which describes the algorithm EcoBeaker uses to run models. Here I give more brief explanations, and tell you how to make changes.

Skipping Grid-Based Species

For many models, the Grid-Based species have no function other than to settle and just sit there. In this case, you may want to do only settlement procedures for Grid-Based species, but not worry about their transition matrices or action procedures. Skipping the transition matrices and action procedures of Grid-Based species can greatly speed up a model, since this means that the simulator does not need to go through every single square in the grid during each time step.

To have EcoBeaker skip Grid-Based species:

- Select the OTHER... item from the Setup menu.

- The Other Setup box will appear. In the top left of this dialog box is a check-box labeled Skip Grid. Check this box to have EcoBeaker skip Grid-Based species. Uncheck it to have EcoBeaker process Grid-Based species.

- Click on the ok button.

When you have Skip Grid checked, EcoBeaker will still do Settlement procedures for Grid-Based species, and will do everything for Individualistic species as normal. You can switch between skipping the Grid and not skipping the Grid while a model is running without the model being reset.

Randomizing the Order of Events

The order in which things happen in an individual-based model can greatly affect the outcome. For instance, imagine a situation in which creatures from two different species are competing for food. You want to see which species is better adapted and will win out. If the creatures of one species always act before the creatures of another species each timestep, this order will give the first species an inherent advantage that has nothing to do with the way you set up the model.

Obviously you don't want this to happen, and there are a couple ways to get around it. One is to use two grids, as described below under the "Flipping Grid" section. That has problems of its own, however. Another way around the problem is to randomize the order in which creatures act each timestep. This way, a creature that had an advantage by going first in one timestep may end up going last in another timestep, and so, on average, no single creature will have any advantage. This does not completely overcome the problem, since in a given timestep some creatures will still have the advantage of going first (or last, or wherever is better), but for most practical purposes randomizing works well.

Randomizing each timestep can be slow. EcoBeaker uses a pretty fast randomization algorithm, but its use still slows down the models noticeably. Because of this effect, EcoBeaker allows you to choose how much randomization you want.

The order in which creatures of Individualistic species act is always randomized each timestep, and there is no way to turn this off. However, the order in which the program goes through the squares on the Grid doesn't need to be randomized. The default is to go through the Grid starting at the upper-left square, down the first column, down the second column, and so on, until the lower-right square, in each square performing the transition matrix and action procedures for Grid-Based creatures. This will happen before Individualistic creatures act. You can randomize the order in which EcoBeaker goes through the Grid squares, still having all Grid squares processed before any Individualistic creatures are processed. You can also have EcoBeaker randomize the Individualistic and Grid-Based creatures together, so the processing of Grid-Based and Individualistic species is mixed together.

To change the randomization algorithm that EcoBeaker is using:

- Select the OTHER... item from the Setup menu.

The Other Setup box will appear. In the upper-left of this dialog box is a set of three radio buttons that control the randomization algorithm.

- Click on the RANDOMIZE GRID radio button to randomize the order in which Grid squares are processed but to keep all Grid squares processed before Individualistic creatures.

- Click on the FULLY RANDOMIZE radio button to randomize the Grid squares and mix their processing together with that of Individualistic creatures.

- Click on the RANDOMIZE INDIV ONLY radio button to randomize only Individualistic species and not randomize the processing of the Grid squares.

- Click on the OK button.

If that is confusing to you, but you are interested in the randomization, you may want to read the "How EcoBeaker Works" section of this manual, which explains the modeling algorithm of EcoBeaker. Then come back and reread this section.

Note that if Skip Grid is on, this whole discussion of randomization is moot, since the grid squares are ignored.

Flipping Grid

The default modeling algorithm in EcoBeaker works with a single grid. Everything that happens affects this grid at the time it happens. Since creatures act one at a time, this means that each creature that acts during a timestep sees a slightly different world than the creature before it did. In other words, the actions of all the creatures are not synchronous. This can create several kinds of problems. Many of these problems can be cured by randomizing the order of action, as discussed above.

For some order-of-action problems, however, randomizing won't help. For instance, let's say you have a plant species whose survival rate changes depending on how many other nearby plants there are. Put a clump of three plants of this species onto the grid. One of them will act first, and it will see that it has two neighbors. Let's say that a density of two neighbors gives it a 30% chance of dying, and it loses the roll of the dice and dies. Now the next plant in the clump goes, but now it only has one neighbor. Let's say with only one neighbor it has an 70% chance of dying, and the roll of the dice say it dies too. Finally, the third plant acts, but now it has no neighbors. Let's say with no neighbors there is a 100% chance of dying, so this last one dies also.

In this example, all three individuals in the clump died, even though originally each of them only had a 30% chance of dying, so by rights, only one of them should be dead. We got this result because the plants didn't all act at the same time, as they would in real life. Randomization would not help at all in this case.

One way around this problem is to use something called a Flipping Grid. Instead of just one grid, you have two grids. When a creature acts, it looks around in the first grid, and then puts the result of its action into the second grid. At the end of the timestep, you flip grids, so that in the next timestep creatures act based on the second grid, and put the results of their actions into the first grid.

In the example above, the first plant in the clump would die, but it would record its death in the second grid, not erasing itself from the first grid. So when the second plant went, it would still see two neighbors, even though one of them has already decided that it's going to die. The death doesn't actually occur until each plant has acted.

There is one major problem with having flipping grids. If you have a creature in the world that affects what happens anywhere except in its own square—for instance, if it moves or kills things in neighboring squares—then there is conflict-resolution problem. Imagine two rabbits on either side of a piece of grass, in a model using flipping grids. Suppose one of the rabbits acts first, and goes and eats the grass. However, the fact that it ate the grass is recorded in the second grid, but not in the first. Now the second rabbit moves, and it sees that the grass is still there (because it's looking in the first grid). So it, too, goes and eats the grass, and records this in the second grid. In the end, both rabbits have eaten the same blade of grass. This violates the laws of physics, not to mention being ecologically unreasonable. You could imagine a whole population of rabbits sustained this way on single blades of grass. (Maybe that's why they can "breed like rabbits".)

I have been unsuccessful in designing a partially flipping grid that gets around this problem. Thus, for now, (until I figure out a better way to do a flipping grid), you should not use Individualistic species with a flipping grid.

EcoBeaker's Flipping Grid algorithm flips grids exactly as I described. There are two grids, and the creatures act according to one of them, and then put the results of their actions in the other. At the end of the timestep, the two grids are flipped.

EcoBeaker defaults to using a non-flipping grid. To have EcoBeaker use a flipping grid:

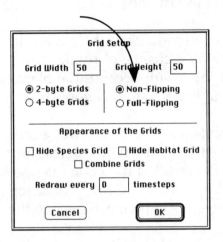

- Select the GRID... item from the Setup menu. This will bring up the Grid Setup box.

- Near the top right of the Grid Setup box are two radio buttons that say "Non-flipping" and "Full-flipping". Click on the FULL-FLIPPING button to have EcoBeaker use a flipping grid. Click on the NON-FLIPPING button to switch back to a non-flipping grid.

- Click on the OK button.

You can change between these while a model is running, but you should not switch to full-flipping if you have Individualistic species in your model.

Graphs

While in many cases you can get an idea of what's going on in a model simply by looking at the patterns in the Species Grid, it is also useful to summarize these patterns. As scientists, we usually summarize data by measuring a quantity or statistic, and putting it into either a graph or a table. EcoBeaker includes many statistics that can be displayed in several types of graphs and tables. The statistics currently included are listed in the "Statistics" section of this manual, with an explanation of each. In this section I describe the different types of graphs and tables available, and how to set up each of them.

Types of Graphs

Currently there are four types of graphs and two types of tables available. These are: line graphs, bar graphs, histograms, and X-Y graphs. The two tables are time tables and stat tables.

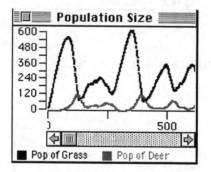

Line graphs show how some variable(s) change over time. The *x*-axis shows time.

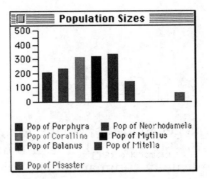

Bar graphs show the value of a series of statistics graphically as the heights of bars.

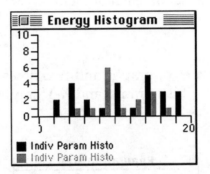

Histograms show distributions of some statistic(s), and can be plotted with either bars or lines.

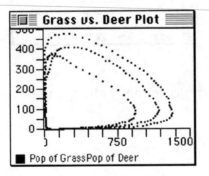

X-Y graphs show two statistics plotted against each other.

```
┌─────────────────────────────────────┐
│ ▤□▤▤▤▤▤▤ Table ▤▤▤▤▤▤▤▤▤▤            │
├─────────────────────────────────────┤
│ Time  Pop of Grass  Pop of Deer     │
│ 721   497           31         ⬆     │
│ 722   492           32               │
│ 723   488           33               │
│ 724   490           35               │
│ 725   491           37         ▤     │
│ 726   491           39               │
│ 727   488           39               │
│ 728   481           43         ⬇     │
│ 729   475           43         ▣     │
└─────────────────────────────────────┘
```

Time tables are like line graphs in that they give the values of some variables over time, but they display the values in a table of numbers.

```
┌─────────────────────────────────────┐
│ ▤□▤▤▤▤▤ Indices ▤▤▤▤▤▤▤             │
├──────────────────────────┬──────────┤
│ Lloyds of Corallina      │ 1.2071   │
│ Simpson Diversity Index  │ 7.2027   │
└──────────────────────────┴──────────┘
```

Stat tables show the value of certain variables at the current timestep, as numbers instead of in a graph format.

Making a New Graph or Table

To make a new graph or table, do the following:

- Select the GRAPHS... item in the Setup menu. This will bring up the Graph Setup window.

The Graph Setup window is very similar to the Species and Habitat Setup windows. On the right is a list of all the currently open graphs and tables. On the left is a series of buttons that you can use to make, modify, and delete graphs and tables.

- Click on the ADD GRAPH button in the left side of the Graph Setup window.

- A dialog box will come up asking you to pick a type of graph or table. Click on the button for the type of graph you want to make.

Once you select the type of graph you want, the Graph Setup box will appear. This is the main dialog box for designing graphs, where you can add items to the graph or table, and set its characteristics. Each part of this setup box is explained in a section below. Here's a picture of the Graph Setup box:

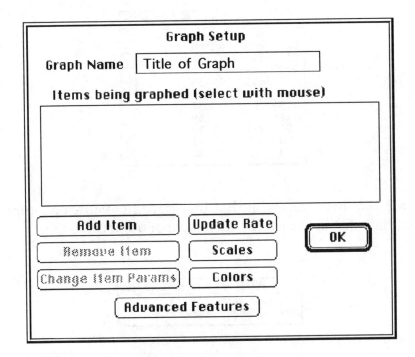

Modifying a Graph or Table

To modify a graph or table, double-click anywhere within the graph's window. This will bring up the Graph Setup box. You can then change any overall characteristics of the graph as described in the sections below. You can also bring up the Graph Setup box for a graph by double-clicking on the graph's name in the Graph Setup window, or by selecting the name and clicking on the MODIFY GRAPH button.

To change one of the statistics in the graph, first bring up its Graph Setup box, then double-click on the name of the statistic. Alternatively, you can select the statistic by clicking once on its name and then click on the CHANGE ITEM PARAMS button. Either of these methods will bring up the Statistic Setup box for that statistic, which you can change as described next, in the "Adding Statistics to a Graph or Table" section of this manual.

Adding Statistics to a Graph or Table

The graphs and tables in EcoBeaker are organized as a series of statistics. Each statistic will be plotted separately on the graph. They range from simple statistics about a single species, such as the species population size, to complicated statistics such as an index of biodiversity. You can have up to ten statistics plotted in each graph.

To add a statistic to a graph:

- Bring up the Graph Setup box for the graph (see "Modifying a Graph or Table" above).

- Click on the ADD ITEM button in the Graph Setup box. This will bring up the Statistic Setup box.

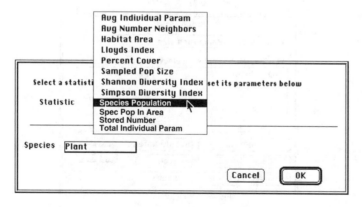

The Statistic Setup box lets you select a statistic and then set parameters for that statistic. The default statistic is Population Size (for all graphs but histograms), and this statistic has a single parameter—the species whose population size you want to know.

- To change the statistic used, click on the name of the current statistic and hold down the mouse button. A menu will pop up with all the currently available statistics. Descriptions of each of these are given in the Statistics section of this manual. Select the one you want, then let go of the mouse button.

- After you select the statistic you want, any parameters that you need to set for that statistic will appear in the bottom half of the dialog box. Set the parameters the way you want them (see the Types of Parameters section for instructions on how to set each kind of parameter).

- Click on the OK button.

A name will automatically be selected for the statistic, and this name will appear in the Graph Setup box.

Note that histograms have a different set of statistics available than do the other graphs and tables, because they need a distribution of data instead of a single data point.

Note also that for X-Y graphs, you must have an even number of items. The items in the left column will be put on the *x*-axis, and will be plotted against the item next to them in the right column.

Removing Statistics from a Graph or Table

To remove a statistic from a graph or table:

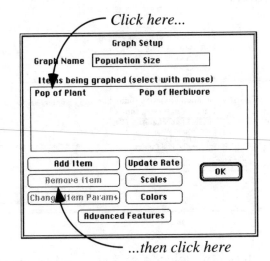

- Bring up the Graph Setup box for the graph (see Modifying a Graph or Table above).

- Find the name of the statistic that you want to remove in the upper part of the Graph Setup box, and click once on the name to select it (it should be highlighted).

- Click on the REMOVE ITEM button. The statistic should now disappear from the setup box.

- Click on the OK button.

The information item you selected will be deleted, and all the other information items will be moved up to take its place.

Setting the Scales of a Graph

The *y* axis in all EcoBeaker graphs can either go from 0 to *y*, or from –*y* to +*y*, where *y* is any positive number. This scale is set by specifying the top of the scale, *y*, which is the maximum value that can be plotted on the graph, and also by specifying whether the *y* axis goes negative or not. You can have all the information items in a graph plotted on the same scale, or you can have individual scales for each item. You can also have EcoBeaker automatically set the scales of the graph, based on the data that is being plotted.

For line graphs, you can also set the length of time plotted across the *x*-axis.

For X-Y graphs, the *y*-axis scale determines the scale of both axes. If you want a different *x* and *y* scale, you can use Separate Maximums (see "Setting the Y-Axis" below).

To set either the *x* or *y*-axis scales, you must first bring up the Scales box for the graph as follows:

- Bring up the Graph Setup box for the graph (see Modifying a Graph or Table above).

- Click on the SCALES button in the Graph Setup box.

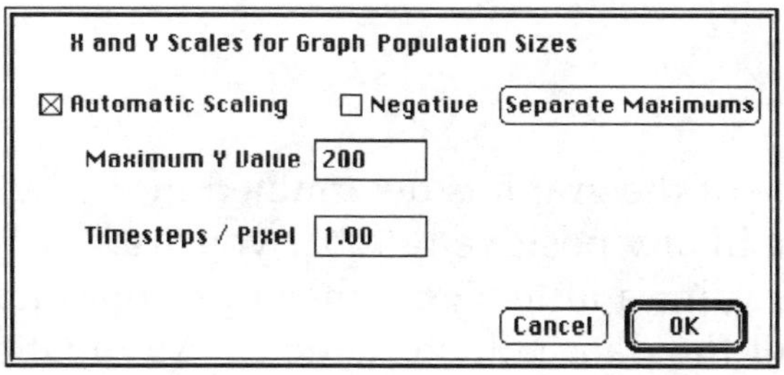

The Graph Scales Setup box should now appear. The next two sections describe what you can do from the Scales Setup box.

Note that there are no scales for Tables or Statistics windows.

Setting the Y-Axis Scale

To set the scale used for the y-axis of any graph, you must first bring up the Scales dialog box, as described just above. The top items in this dialog box determine how the graph's y-axis will look. This includes whether the y-axis is automatically scaled, whether it goes below zero, whether there are separate y scales for each statistic or a single scale, and a place where you can input the maximum y value.

Automatic Scaling

The Automatic Scaling check-box tells EcoBeaker whether you want to set the y-scale of the graph, or whether you want the program to do it automatically as the data come in. If Automatic Scaling is checked, then each time the graph is updated, if any of the data is more than the maximum y value of the graph, the maximum y value of the graph will be increased to twice the value of the data. If there is a negative data point, automatic scaling will make the y-axis go negative. In this way, all data points will always show up on the graph. If Automatic Scaling is not checked, then whatever maximum you set will be the maximum value of the graph, and any data points that fall higher than this will not show up on the graph (in the case of line graphs) or will make a bar that extends to the top of the window (for bar graphs). The one drawback to automatic scaling is that it will never decrease the y scale if the data values get smaller.

Negative Values

The Negative check-box determines whether negative values will show up on the graph. If Negative is checked, then the graph will go from $-y$ to $+y$, where y is the maximum value that is shown on the graph. Otherwise, the graph will go from 0 to y.

Maximum Y Value

The maximum y value of the graph is determined in the field labeled Maximum Y Value. You can type in any positive number you want here. It's a good idea to make the maximum y value a little bigger than the largest number you think will get plotted, so that all the data will show up. If Automatic Scaling is checked, then this maximum y value is only the initial maximum, and if any data goes above it, then the maximum y value is increased, as discussed above.

Separate Maximum Y Values

On the top at the right of the Scales Setup box is a button labeled Separate Maximums. The default for graphs in EcoBeaker is to have one maximum *y* value for all the statistics in the graph. However, you can make a separate *y* scale for each statistic by clicking on the SEPARATE MAXIMUMS button. When you click on this button, the field where you input a maximum *y* value will be expanded into ten fields to input maximum *y* values, one for each of the possible statistics in this graph. The names of the currently included statistics will be printed next to the appropriate input fields. Put the maximum value you want for each statistic into the accompanying input field.

When you are using separate *y* scales for each statistic, the SEPARATE MAXIMUMS button will be changed into a ONE MAXIMUM Y button. If you want to change back to a single maximum, click on the ONE MAXIMUM Y button, and the dialog box will revert to its original state, with only one maximum *y* value.

Setting the Time Axis

For line and X-Y graphs, you can set the scale of the *x*-axis. The *x* scale of an X-Y graph behaves the same as the *y*-scale discussed above, and the Scales Setup box for an X-Y graph will contain a field where you can input the *x* scale.

For line graphs, the *x* scale shows time, and it's specified by telling the graph how many timesteps to plot on each column of pixels across the window. One pixel is one of the little dots on the screen. An average 14" screen is 640 pixels across, so if your graph takes up about a third of the width of the screen, then it's about 200 pixels long. If you want to plot 1000 timesteps in this 200 pixel window, you want the x-axis of the graph to have a scale of 1000 / 200 = 5 for timesteps per pixel. In general, since you won't really know how many pixels wide your window is, you should experiment with different timesteps/pixel until you get one that you like.

To set the *x* scale of a line graph:

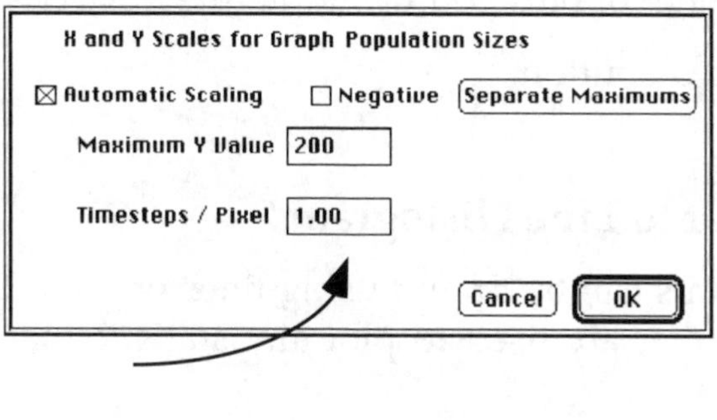

- Bring up the Scales Setup box for the line graph (see "Setting the Scales of a Graph" above).

- At the bottom of the Scales Setup box is a text field labeled Timesteps / Pixel. Type in the number of timesteps you want plotted in each column of pixels.

- Click on the OK button.

Setting the Bin Sizes for Histograms

When you make a histogram, you need to specify the number of bins in the histogram, and the size of each bin. In EcoBeaker, you do this by specifying the start of the first bin in the graph, the end of the last bin, and the total number of bins in-between. EcoBeaker then calculates how big each bin should be. For instance, if start bin is 0, end bin is 10, and number of bins is 10, then there will be ten bins, each 1 unit wide, with the first bin going from 0 to 1, the second from 1 to 2, etc.

To set up the bins of a histogram:

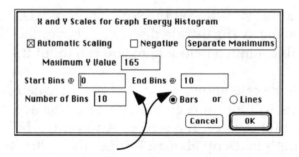

- Bring up the Scales Setup box for the histogram (see "Setting the Scales of a Graph" above).

- At the bottom of the histograms Scales Setup box are three text fields used to set up the bins. Put the beginning value and the ending value of the histogram into the Start Bins and End Bins text fields, respectively.

- Put the number of bins you want into the NUMBER OF BINS text field.

- Click on the OK button.

Changing from Bar to Line Histograms

EcoBeaker's histograms can be drawn using lines or bars. This is useful because histograms in EcoBeaker are used to plot any statistic that produces a distribution, and some of these are easier to look at as line graphs than bar graphs. For instance, the histogram is used to plot autocorrelation, and in that case you would probably want to use lines rather than bars.

The default histogram uses bars to plots its values. To switch a histogram so that it plots with lines:

- Bring up the Scales Setup box for the histogram (see "Setting the Scales of a Graph" above).

- At the lower-right of the histogram's Scales Setup box are two radio buttons labeled Bars and Lines. Click on the LINES button to have the histogram plot using lines. Click on the BARS button to have the histogram plot using bars.

- Click on the OK button.

Setting the Update Frequency of a Graph or Table

Some graphs and tables can take a while to gather the information they need and plot it, and in the interest of speeding up the simulation, you may not want the graph or table to recalculate and redraw its statistics every time step. You may also want to update a graph infrequently if you are running a model for a long time. You can specify the number of time steps between updates for each graph and table. The default update rate is 1, which means that the graph will be updated every single timestep while the model is running. If you change the update rate to 2, then the graph will only be updated every other timestep, change the rate to 3 and it will be updated every third timestep, and so on.

To change the update rate of a graph or table:

- Bring up the Graph Setup box for the graph (see "Modifying a Graph" above).

- Click on the UPDATE RATE button.

- The dialog box that appears contains a single text field labeled Update Every ___ Timesteps. Type in how many timesteps you want between updates to the graph.

- Click on the OK buttons in both dialog boxes.

Setting the Colors of a Graph or Table

EcoBeaker tries to be smart about the colors it uses for each statistic. For instance, if you plot a statistic about a certain species, EcoBeaker will plot the statistic in the species' color. Sometimes you will want to change the colors that a graph is using, though, and you can do this through the Graph Setup box.

To change the color used to plot a statistic in a graph:

- Bring up the Graph Setup box for the graph (see"Modifying a Graph" above).

- Click on the COLORS button.

- A dialog box will appear with the names of each statistic in the graph and a colored square to the right of each name. Find the name of the statistic whose color you want to change, and click once on the square next to it.

- A standard Macintosh color wheel will appear. Pick the color you want.

- Click on the OK buttons in all the dialog boxes.

Setting the Length of Data Stored in a Graph

Line graphs, tables, and X-Y plots all show a series of data points, and they can only store a certain number of data points before running out of memory. When they run out of storage space, they will start overwriting the initial data points. You will see this if, for instance, you try scrolling backwards in a line graph after running a model for a long time. As you are scrolling backwards, you will reach a point where the graph is blank because data points before that point have been overwritten by later data points. You may want to increase the number of data points that a graph can store so that it takes longer for the initial data of the graph to be overwritten. Alternatively, you may want to decrease the amount of data stored in order to save memory.

By default, a graph will store 10,000 timesteps worth of data. To change the number of data points a line graph, table, or X-Y plot can store:

- Bring up the Graph Setup box for the graph (see "Modifying a Graph" above).

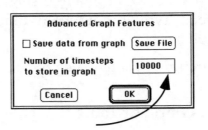

• Click on the ADVANCED FEATURES button. This will bring up the Advanced Graph Features box.

• In the bottom of the Advanced Graph Features box is a text field labeled Number of timesteps to store in graph. Type in the number of timesteps worth of data that you would like the graph to store.

• Click on the OK buttons in both the dialog boxes.

Note that changing this number may reset the graph, so you should do it before starting to run a model.

Saving Graph Data

There are two ways to save the data that is being displayed in a graph. One is to use the Save Data command, which will save all the data in the graph at the current timestep. The second is to have the graph save all the data it plots as the model is running. The first method is described here. The second method is described below under "Saving Graph Data as the Model Runs".

To save the data that you see plotted in a graph or table:

• Select the graph by clicking once in its window (so that the window's title bar is not ghosted).

• Select the SAVE DATA... command from the File menu.

• In the standard Macintosh dialog box that appears, type in the name you wish to give the file. If you type in the name of a file that already exists, EcoBeaker will give you the choice of either overwriting the old file, or appending to it.

• Click on the OK button.

The data will be saved as text, with a separate column for each item plotted in the graph, and the first column giving the timestep. The top two lines in the file will be the name of the graph, and titles for the columns. For stat tables and bar graphs, there will only be one row of data, the data from the current timestep. For line graphs, X-Y graphs, and time tables, there will be a row for each timestep where data is displayed in the graph. See the next section for some subtleties associated with this format. Histograms cannot currently be saved using SAVE DATA, but can be saved as the model runs (see below).

Saving Graph Data as the Model Runs

This section explains how to have the data that is being plotted on a graph simultaneously saved to a file. Alternatively, you can save data from a graph after you have finished running the model. Saving data from a graph after the model has stopped running is described under "Saving Graph Data" above.

There are a couple advantages to saving graph data as the model runs instead of saving it at the end. First, if you save data from bar graphs, histograms, or statistics windows after the fact, you will only save the data for the current timestep, whereas if you save as the model runs, the data will be written to a file each time the graph is updated.

A second advantage is that each line graph, table, or X-Y graph can only hold a certain length of data. If you run a model for long enough, the data points from early in the run will be overwritten by later data points. If you are saving data as you run the model, then all the data points will be saved to the file, but if you save after the run is over, and it was a long run, only the last x timesteps of data will be saved (where x is the length of time for which the graph can store data—see "Setting the Length of Data Stored in a Graph").

To have a graph save data as the model runs:

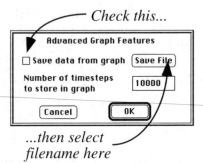

Check this...

...then select filename here

• Bring up the Graph Setup box for the graph (see "Modifying a Graph or Table" above).

• Click on the ADVANCED FEATURES button. This will bring up the Advanced Graph Features box.

• At the top of the Advanced Graph Features box is a check-box labeled Save data from graph and a button called Save File. Check the check-box.

- Click on the SAVE FILE button in the Advanced Graph Features box. This will bring up a standard Macintosh dialog box asking you for a filename. Type in a name for the data file from this graph. If a file with this name already exists, the old file will be deleted.

- Click on the OK buttons in all the dialog boxes.

All the data is saved as text. The first line of the file will give the name of the graph and the name of the model. Under this first line will be one line with labels for each column of data, then the data will follow, with the first column giving the timesteps and each subsequent column containing the data from one of the statistics in that graph. Each time you reset the model, the initial two lines (the title of the graph and the column headings) will be saved to the file again, so that these will serve to delineate the data from different runs of the model.

Histograms have a slightly different data format than other graphs. The first two lines will be the same, but in this case the column headings indicate the titles of the following rows of data. The first row after the column headings will contain a single number, the timestep. After this will be a row of data for each statistic in the histogram, each column of data giving the values in each bin of the histogram. Then the next timestep will be saved, then the next rows of data, and so on. (This format is a very inconvenient way to save the data from histograms, and I'm open to suggestions.)

Note that data is only saved to the file when it is plotted on the graph, so if you set the update frequency to more than one timestep, this will also determine the rate at which data is saved to the file.

Saving and Copying Pictures of a Graph

You may want to export a picture of a graph or table to another program (e.g., a drawing program), where you can touch it up a bit. EcoBeaker lets you save a picture of any graph or table, or copy the picture to the clipboard.

To save a file with a picture of a graph or table:

- Select the graph by clicking once in its window (so that the window's title bar is not ghosted).

- Select the SAVE PICTURE... command from the File menu.

- In the standard Macintosh dialog box that appears, type in the name you want to give the file.

- Click on the OK button.

A picture of the graph will be now be saved in PICT format into the file you specified. You can then load this file into any drawing program for further work.

To copy a picture of a graph or table to the clipboard:

- Select the graph by clicking once in its window (so that the window's title bar is not ghosted).

- Select the COPY command from the Edit menu.

A picture of the graph will be put into the clipboard, and you can now PASTE this picture into another program.

Removing a Graph

To get rid of a graph you no longer want, you may either click in the CLOSE box of the window containing the graph, or select the graph in the Graph Setup window by clicking on it once to highlight it. Then click on the DELETE GRAPH button.

Shrinking, Hiding, and Showing the Graph Setup Window

To avoid clutter on the screen, you can shrink the Graph Setup window, or make it disappear altogether. To shrink it, click on the little icon in the upper-right corner of the window. Clicking once will cause the window to shrink in size so that only the names of the graphs are showing. Clicking on that icon again will expand the window back out. To make the window disappear entirely, click on the CLOSE icon in the upper-left corner of the window. To make the window reappear, you must go to the Setup menu and select the GRAPH item.

Sampling

EcoBeaker includes many commonly used sampling techniques for sampling the distributions of creatures on the Grid. You can take a sample at any time while running a model, and you can also freely switch between sampling techniques. This lets you examine the operation of a single sampling technique or compare the accuracy of two or more sampling techniques. You can also use sampling techniques to make EcoBeaker laboratories more realistic, forcing the person using the lab to sample the distributions of species in the model instead of having perfect information.

The next few sections explain how to choose and set parameters for the sampling techniques, and describe a few more advanced features of EcoBeaker's sampling algorithm, such as saving the samples to a file on disk and automatically sampling at certain timepoints during the simulation.

Taking a Sample

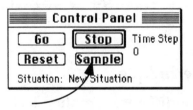

To take a sample, you may either click on the SAMPLE button in the Control Panel, or select the SAMPLE command from the Action menu. Either of these choices will take a sample from the Species Grid using the current sampling technique. The sampling technique will sample only those species you specify, and will show the results of the sample on the screen and save these results to a file. You can also set the sampling to occur automatically at certain timesteps (see "Automatic Sampling" below).

177

The Sample Setup Box

To set anything to do with sampling, select the SAMPLING... item from the Setup menu to bring up the Sampling Setup box. This setup box lets you change the sampling technique, set all the parameters associated with the current sampling technique, and control some advanced features of EcoBeaker's sampling techniques; for instance, you can set up the sampling techniques to automatically take samples at specified times.

Here's a picture of the Sampling Setup box:

Setting the Sampling Technique

EcoBeaker includes several different sampling techniques that are commonly used in ecological studies. Descriptions of each of EcoBeaker's techniques are given in the "Sampling Techniques" section of this manual. You can use only one technique at a time, but you can switch the current sampling technique at any time, either before or while a model is running.

The default sampling technique is NONE. To change the current sampling technique:

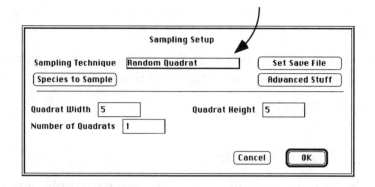

- Select the SAMPLING... item from the Setup menu to bring up the Sampling Setup box.

- At the top of the Sampling Setup box is a pop-up menu labeled Sampling Technique. Click and hold down the mouse button inside of this menu so that the menu pops up.

- Use the mouse to select the sampling technique you want, and then let go of the mouse button.

When you have selected a new sampling technique, the bottom part of the Sampling Setup box will expand to show the parameters for this sampling technique. These are all initially set to default values, which you will probably want to change. The next section gives instructions for changing these parameters.

Changing the Parameters of a Sampling Technique

Each Sampling Technique in EcoBeaker has a set of parameters that control its function. The parameters for each technique are described in the "Sampling Techniques" section of this manual. To change the values of the current techniques parameters:

- Select the SAMPLING... item from the Setup menu to bring up the Sampling Setup box.

- At the bottom of the Sampling Setup box are all the parameters for the current Sampling Technique. Set each of these as you wish.

- Click on the OK button.

Selecting the Species to Sample

Most sampling techniques in EcoBeaker allow you to sample one or more of the species in the model at the same time. To specify which species you want to have sampled:

- Select the SAMPLING... item from the Setup menu to bring up the Sampling Setup box.

- Click on the SPECIES TO SAMPLE button, which is near the upper-left of the Sampling Setup box. This will bring up a new dialog box with a list of all the species currently in the model.

- Check the boxes next to each species that you want to have sampled, and uncheck the boxes next to species you don't want sampled.

- Click on the OK buttons in both dialog boxes.

Samples File

In addition to being displayed on the screen, all samples taken within EcoBeaker are saved to a file so that they can be read into another program for further analysis. Each sampling technique saves its data in a unique format suitable to the data. However, some things are common across sampling techniques. For each sample, there is usually an initial line that specifies the sampling technique and the parameters being used, then a line giving titles for each of the columns in the data to follow. Following the labels are several rows of data, one row for each piece of the sampling. For instance, if you're using quadrats, there would be one row for each quadrat. All of this is pure text, so you can read the file into any spreadsheet or statistics program, or even a word processor.

Naming the File Where Samples Are Saved

The default name of the sampling file is Sampling File. To change this name:

- Select the SAMPLING... item from the Setup menu to bring up the Sampling Setup box.

- Click on the SET SAVE FILE button in the top right of the Sampling Setup box.

- A standard Macintosh dialog box will appear asking you for a filename. Type in the name of the file where you want samples to be saved.

- Click on the OK buttons in both dialog boxes.

Sampling Verbosity

EcoBeaker's sampler lets you set how verbose it is. The verbosity of the sampling determines how much interaction the sampling technique will have with the user. If you set verbosity to INTERACT, the user is asked lots of questions about the sampling, and all the results appear on the screen. If you set it to SHOW RESULTS, the user is asked no questions, but the results appear on the screen. If you set it to NO FEEDBACK, the sampling occurs and is saved to the sampling file (see above), but the results are not shown on the screen. You might want to use this last setting with automatic sampling (described below) if you want to let the model run for some time and be sampled every now and then, but you don't want to sit at the computer and click on OK buttons every time a sample is taken.

To set the verbosity used when sampling:

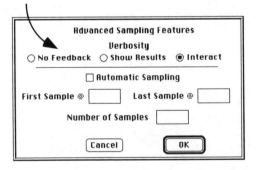

- Select the SAMPLING... item from the Setup menu to bring up the Sampling Setup box.

- Click on the ADVANCED FEATURES button at the bottom of the dialog box to bring up the Advanced Sampling Features box.

- At the top of this second dialog box are three radio buttons labeled No Feedback, Show Results, and Interact. Click on the one corresponding to the level of verbosity you want.

- Click on the OK buttons in both dialog boxes.

Automatic Sampling

Automatic sampling tells the program to automatically take a sample at fixed intervals of time as a model is running. You might do this if you want to take samples at very precise times as a model is running. Alternatively, you might do this if you are running a model for a long time and would like samples during that period, but don't want to sit in front of the computer taking samples by hand.

The timing of automatic samples is set by specifying the time of the first sample, the time of the last sample, and the total number of samples you want to take. The total number of samples is then evenly distributed between the first and last samples. If the total number of samples you request is 1, then a single sample will be taken at the first sampling time.

If you enable automatic sampling by checking the Automatic Sampling checkbox, the program will automatically take Number of Samples samples, the first one taken at First Sample @, the last one at Last Sample @, and the rest at regularly spaced intervals in between. Depending on your reason for doing this, you may want to turn verbosity to No Feedback.

To turn on and set up automatic sampling:

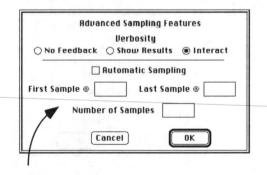

- Select the SAMPLING... item from the Setup menu to bring up the Sampling Setup box.

- Click on the ADVANCED FEATURES button at the bottom of the dialog box to bring up the Advanced Sampling Features box. At the bottom of this dialog box are the items for automatic sampling.

- Click on the AUTOMATIC SAMPLING check-box so that its checked.

- Enter the timestep at which you want the first sample taken in the field called First Sample @, the time you want the last sample taken in the field called Last Sample @, and the total number of samples you want taken in the field called Number of Samples.

- Click on the OK buttons in both dialog boxes.

Marking Interesting Runs of a Model

Sometimes you may run a model for a while and reach a point that you find interesting. You may want to save this run of the model so you can compare it to later runs. One way to do this is to save this run as you would save any model, using the SAVE SIMULATION command described above. If you tell EcoBeaker to include data in the file, you will then be able to reload that file and get back to exactly this same position in the model.

Using the SAVE command in this situation is a bit clumsy, however, and EcoBeaker has a second, more convenient mechanism to achieve the same effect. You can set a Marker at the point in the model that you want to save, then come back to the marked position whenever you choose.

You can use this capability in many ways. One way is to compare between different runs of a model. Let's say you want to run a model ten times for a given number of timesteps, and then visually compare the patterns you see on the screen at the end of each run. To do this, run the model once, and when you reach the stopping point, set a marker called Run 1. Then reset the model, run it again, and at the end set a marker called Run 2. Keep doing this through Run 10. Now go to the Markers menu and select the run you want to view. The model will return to the state it was when you set the marker.

Along similar lines, you may want to run a model for a long time, and compare the way the model looked at different time points along the way. Or you may reach an interesting point in a run of a model and want to run the model multiple times from that point. I do this for laboratories in which it takes a while for the model to reach an equilibrium and I want students to start experimenting from the equilibrium point without sitting through the startup time.

When you set a marker, the internal procedure used by EcoBeaker to save the model is exactly the same as that used by the SAVE SITUATION command. Not only is all the data from the model saved, but other important information is also saved—the positions of all the windows in the model, the list showing which species are in the model, and so on. Thus, you can use the markers as an easy way of switching between different models if you will be going back and forth regularly. For instance, you may want to write a lab in which you ask your students to try working with more than one model.

Unfortunately, at this time there is no way to view two different marked positions side-by-side on the screen. You have to switch back and forth and view them one at a time. You can, of course, copy the data you want to view side-by-side to a drawing program.

Setting a Marker

To set a marker, go to the Markers menu and select the ADD MARKER item. You will then be asked to name this marked position. Give it a name and click on OK.

Now go back to the Markers menu. At the bottom of the menu you will see the name you just gave to this position in the model. You can get back to that position at any time by selecting its name from the Markers menu.

Resetting the Simulation to a Marked Run

In some models, when you press the RESET button you don't want the Grids to become completely empty and the timestep to go back to 0. Instead, you have a particular point in the model that you would like to reset to. Perhaps you have set up a landscape with a bunch of creatures in it, and you want that to be the starting point. If you have saved this point in the model as a Marker (described above), you can instruct EcoBeaker to reset the model to this marker instead of resetting to time 0.

To have a model reset to a marker, do the following:

- Add a Marker that stores the position you want to reset the model to (see the "Setting a Marker" section above).

- If you have run the model since you added the Marker, select the Marker from the Markers menu. The model should now be set up exactly as you want it to be when it gets reset.

- Select the OTHER... item from the Setup menu. This will bring up the Other Setup box.

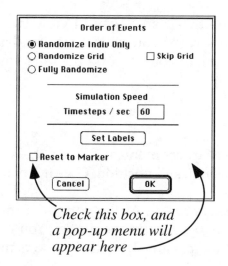

*Check this box, and
a pop-up menu will
appear here*

• At the bottom of the Other Setup box is a check-box labeled Reset to Marker. Check this box. (This check-box won't appear unless you have already set at least one Marker).

• When you check RESET TO MARKER, a pop-up menu containing the names of all the markers currently saved will appear next to it. Click and hold down the mouse button in the menu to pop it up.

• Move the mouse to select the Marker you want from the pop-up menu, and then let go of the mouse button.

• Click on the OK button.

• Select the ADD MARKER... command from the Marker menu to add a marker, and give this new marker exactly the same name as the marker you selected from the pop-up menu above.

• A dialog box will appear saying that you already have a Marker of this name, and asking if you want to replace it. Click on the YES button.

You must do these last two steps so that the marker knows it's supposed to reset to itself. Now any time you reset the model using the RESET command, it will reset itself to the Marker you specified. To have EcoBeaker reset the model back to time 0, simply uncheck the RESET TO MARKER check-box in the Other Setup box as described above.

Parameter Window

I have tried to make the user interface for EcoBeaker as simple as possible, so that you can play with models with a minimum of effort in learning how to use the program. However, the normal method of changing parameters in EcoBeaker still involves working with one or more dialog boxes. The labels for parameters can also be somewhat cryptic since they are space-limited. Thus I have added in an optional window—the Parameter window—which lets you bypass the dialog boxes for selected parameters.

The Parameter window allows you to place parameters which you or others may want to change, in a window on the screen. It also lets you label each parameter in this window with a full sentence. You can display any species, graph, habitat, or sampling parameter in the Parameter window, and the window can hold up to ten separate parameters.

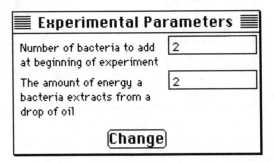

A Sample Parameter Window

There is one more thing that the Parameter window is useful for. Sometimes you may wish several parameters in the model to always be the same. You can do this by attaching them together in the Parameter window.

Using the Parameter Window

To change a parameter displayed in the Parameter window, first locate the parameter in the window. Then simply type in a new number or select the new value for the parameter. When you have set the new value, click on the CHANGE button at the bottom of the Parameter window. You can also change the values by resetting the model, as long as you are not resetting to a saved marker (see the "Action Commands" section of this manual). The parameter will not be changed until you click on the CHANGE button or reset the model.

Making the Parameter Window

Making a Parameter window is a several-step process. You must first make the window and give it a title, then add the parameters you want, and finally, size the window properly. You may also want to attach certain parameters together, so they get changed in concert.

To make a Parameter window:

• Select the PARAMETER WINDOW... item from the Setup menu. This will bring up the Parameter Window Setup box.

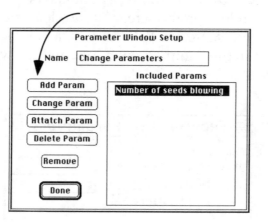

• To add a parameter to the Parameter window, click on the ADD PARAM button in the setup box.

• A new dialog box will appear, asking if you want to add a parameter from a species, a habitat, a graph, or the sampling procedure. If there are no habitats or graphs in the model, those options will not be available. Click on the appropriate button.

Now the Parameter Selection box will appear, which is where you actually select the parameter. For species and graphs, there are three things you must set to select a specific parameter. First, from a species or graph, choose the parameter you want to add. For species, you must then select whether you want a parameter from the Settlement procedure, the Action procedure, or the Transition Matrix. For graphs, you must specify the item in the graph from which you want a parameter. Finally, you must select the parameter.

These three things are selected from three pop-up menus in the Parameter Selection box. The top pop-up menu will give the species or graph, the middle menu gives the procedure or item, and the bottom menu gives the actual parameter.

- Select the items in the Parameter Selection box, in order, from top to bottom: first the species or graph, then the procedure, and finally the parameter. This order is shown in the diagram below:

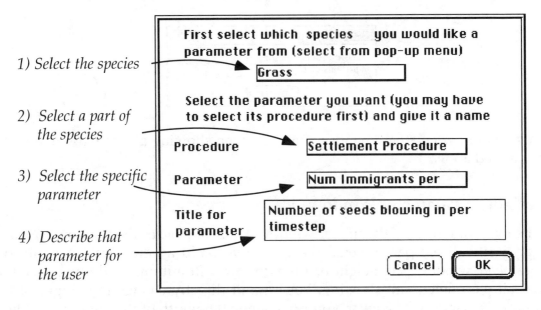

The Parameter Selection Box

This figure shows an example of selecting a parameter for inclusion in the parameter window. The user has selected the Num Immigrants per Turn parameter of the Settlement Procedure of a species called Grass. The instructions on the left show the order in which you should select each item.

For parameters from a habitat, only the first and last pop-up menu matter, since there is only one procedure in a habitat. For parameters of the sampling procedure, only the last pop-up menu matters. If there are no parameters for a given procedure, then the third pop-up menu will say No Parameters, and you cannot select anything here.

- At the bottom of the Parameter Selection box is a text field in which you can type in a title for the parameter. This will become the label for that parameter in the Parameter window. Write a title that describes how the parameter is used in your model.

- Click on OK button in the Parameter Selection box. That parameter will now be added into the Parameter window.

- Repeat the above steps to add any more parameters you want to have in your Parameter window.

- At the top of the Parameter Window Setup box is a text field labeled Name. Type in a name for your Parameter window.

- Click on the OK button in the Parameter Window Setup box.

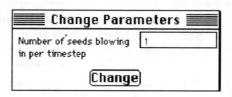

There will now be a new window on the screen showing all the parameters that you added above.

Resizing a Parameter Window

When you first make a Parameter window, EcoBeaker sizes the window so that it exactly fits all the parameters. Every time you add a new parameter into the Parameter window, the height of the window will automatically be resized so that all parameters are showing. You cannot directly change the height of the Parameter window. However, you can change the width of the window by clicking in the lower-right corner of the window and dragging it around, just as you would with any standard Macintosh window. There is a size box there, even though it's not drawn. EcoBeaker will then resize the height of the window so that everything just fits, leaving the width as you set it.

Attaching Parameters Together

In some instances you may want one parameter in the Parameter window to change several other parameters in the model. For instance, if you have two species you want to always move at the same speed, then you might want one item in the parameter window labeled Speed, which will change the speed of both species when the user modifies it. Another example is in the Island Biogeography lab, where there is an island in a certain position in the grid, and the population graph needs to show the population sizes only on the island. When the user changes the position of the island, the area being sampled for the graph should also change.

EcoBeaker includes a way of attaching parameters together. To attach several parameters together:

- Select the PARAMETER WINDOW... item from the Setup menu to bring up the Parameter Window Setup box.

- If you want to attach several parameters together, one of them must be the main parameter. Add this main parameter as described above in Making the Parameter Window.

- Click once on the main parameter so it is selected.

- Click on the ATTACH PARAM button in the Parameter Window Setup box.

- The Parameter Selection box will appear. This is the same dialog box you used to select the main parameter. In this dialog box, select the parameter you want attached to the main parameter.

- Repeat the last two steps until you have selected all the parameters you want attached together.

- When you are finished, click on the DONE button in the Parameter Window Setup box.

Now whenever you change the value of these parameters in the Parameter window, the values of all the parameters you attached together will be changed.

You can attach as many parameters together as you want. The only restriction is that all the parameters must be of the same type as the main one (i.e., you can only attach number parameters to number parameters, species parameters to other species parameters, and so on). Also, keep in mind that there is no way to see the attached parameters, and no way to modify which parameters are attached to a main parameter. So if you make a mistake, you'll have to delete the main parameter and start over again. Whenever you delete the main parameter, all the attachments are deleted as well.

One last note: The parameters are only attached together if their values are changed through the Parameter window. If you change only one of these parameters by the normal means of changing parameters—through the species setup box, the sampling setup box, and so on—the other parameters attached to this one will be unaffected.

Modifying the Parameter Window

After you have set up a Parameter window, you may want to modify or remove one of the parameters. To modify the Parameter window:

- Select the PARAMETER WINDOW... item in the Setup menu to bring up the Parameter Window Setup box.

- In the Parameter Window Setup box, click once on the parameter you wish to modify or remove. The parameter should now be highlighted.

- To modify the parameter, or to change its title, click on the CHANGE PARAM button. This will bring up the Parameter Selection box, and you can follow the instructions in "Making the Parameter Window" to make changes.

- To remove this parameter from the Parameter window, click on the REMOVE PARAM button.

- Click on the OK button.

After you make any changes, the height of the Parameter window will automatically be resized to fit the remaining parameters.

Note that EcoBeaker will automatically remove parameters when they no longer exist in the program. For instance, if you have a parameter showing in the Parameter window from a species that you then delete, that parameter will also be removed from the Parameter window. This is to prevent problems that might arise from trying to modify a parameter that isn't there anymore. However, this mechanism isn't perfect, so you should build the parameter window in your models last, after everything else about the model is set.

Getting Rid of the Parameter Window

If you have put a Parameter window into a model and decide you no longer want it, you can get rid of it through the Parameter Window Setup box. To get rid of the Parameter window:

- Select PARAMETER WINDOW from the Setup menu, to bring up the Parameter Window Setup box.

- Click on the REMOVE button in the dialog box.

- An alert will appear asking if you really want to get rid of the Parameter window. Click on the YES button.

The Parameter window will now disappear.

Odds and Ends

This section includes additional features of EcoBeaker that didn't really fit else-where in the manual, and aren't big enough topics to constitute a section in and of themselves. Some of the features discussed here include

- Manually adding creatures to the grid
- Viewing individuals on the grid
- Renaming windows and phrases
- Reducing the speed of models
- Memory management issues

Painting Mode

You may want to get into the Species Grid or Habitat Grid while a model is run-ning and change settings, erase creatures from some squares, add them into oth-ers, and generally create havoc in the world. You can do this by switching into Painting Mode. Painting Mode is a mode in EcoBeaker in which you can use the mouse to add creatures onto the Species Grid, or habitat onto the Habitat Grid.

To use Painting mode:

- Stop the model from running (see the "Action Commands" section of this manual).

```
╔══════ Control Panel ══════╗
║ Painting    [ Stop ]  Time Step ║
║ Mode                  0          ║
║                                  ║
║ Situation: Example Situation     ║
╚══════════════════════════════════╝
```

- Select the PAINT command from the Action menu. You will see a mes-sage in the Control Panel telling you that you are in Painting mode.

• Find the Species Setup window and click once on the name of the species that you want to add to the Species Grid, so that species is highlighted. For instance, if you want to make some of the squares in the Species Grid Empty, then go to the Species Setup window and click once on the species Empty.

• Click once anywhere within the Species Grid window to make that the active window (its title bar should not be ghosted).

• Move the mouse over the position in the Species Grid where you want to add an individual of the species you selected above.

• Click the mouse button once to add one individual of the species you selected at that grid square. The creature that was at that square will now be replaced by a creature from the currently selected species.

• If you want to add a rectangular patch of creatures, move the mouse to one corner of the rectangle, push and hold down the mouse button, and drag the mouse towards the opposite corner. As you move the mouse, you'll see a rectangle outlined on the grid. When you let go of the button, this entire rectangle will be painted with the species you selected.

• To paint habitats, you follow a very similar procedure as outlined above for species. Select the habitat you want to paint from the Habitat Setup window, and then use the mouse to paint the new habitat onto the Habitat Grid.

• When you are done painting, leave Painting mode by clicking on the STOP button in the Control Panel, or by going to the Action menu and selecting PAINT. Either way, the Control Panel will now return to normal, indicating that you are back in the regular mode.

Getting Info About Creatures

You may want to explore what the creatures in the current model are like by looking at individual creatures. This information is particularly useful if you have Individualistic species in the model and you want to look at the individual parameters of creatures in the grid. You may also want to know the exact position of a particular creature in the grid.

To look at a creature, move the mouse to point at that creature, then press and hold down the mouse button. A window will appear showing information about the creature. At the top of the window will be the name of the species to which that creature belongs, and its (*x*, *y*) location on the grid. If it's a member of an individualistic species, then its individual parameters will appear below the species name.

If you want to look at several creatures, you can either click on each one separately, or you can just hold the mouse button down and move the mouse from one to another. As you move the mouse around, the information in the info window will change so you always see what's under the pointer.

Tagging Creatures

In some models, where lots of things are moving around, it may make it easier for you to see what's going on if you can follow a single individual as it moves. To facilitate this, EcoBeaker lets you mark an individual with a special color. That individual will then be drawn in the color you selected instead of the normal color for its species. EcoBeaker calls this "Tagging" the individual. You can only tag creatures from Individualistic species.

To tag a creature:

> • Stop the model from running (see the "Action Commands" section of this manual).

```
╔═══════ Control Panel ═══════╗
║ Tagging   ┌──────┐  Time Step║
║           │ Stop │           ║
║ Mode      └──────┘    0      ║
║                             ║
║ Situation: Example Situation ║
╚═════════════════════════════╝
```

> • Select the TAG command from the Action menu. You will see a message in the Control Panel telling you that you are in Tagging mode.

> • Move the mouse to the Species Grid and click once on the creature whose color you wish to change.

> • A standard Macintosh color requester will appear. Pick the color you want to use for that creature.

> • Click on the OK button.

From now on, that creature will be drawn in the color you selected.

Naming Things

When you write a new model, you may want more descriptive names for the different windows and grids than EcoBeaker normally gives them. For instance, in a model about a forest, labeling the grid window The Forest is much more meaningful than the default title Species Grid. EcoBeaker lets you rename the species and habitat grids; the species and habitat setup windows; Timestep, Settlement Procedure, and Action Procedure.

To rename any of the above items in EcoBeaker:

- Select the OTHER... item from the Setup menu to bring up the Other Setup box.

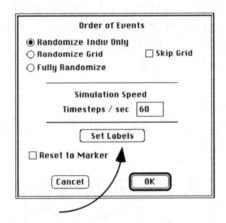

- Click on the SET LABELS button near the bottom of the dialog box. This will bring up a second dialog box containing a list of all the items you can rename in EcoBeaker, with a text field to the right of each item.
- Type in the name you want each item to have.
- Click OK in both dialog boxes.

These new names will be used until you load in a new situation file, and will be saved if you save the situation you're working on.

Slowing Down Time

I have spent a lot of effort trying to make the models in EcoBeaker run as quickly as possible. Complicated models still don't run as fast as I'd like, but some simpler models may run quite quickly, sometimes so quickly that it's hard to follow the action on the screen. Other models will also start out running quickly and then slow down as more and more creatures are added. This can give a student who is not paying attention a distorted sense of time as the model runs. For both of these reasons, EcoBeaker includes a way to slow down the speed at which a model runs.

To slow down the speed at which a model is running:

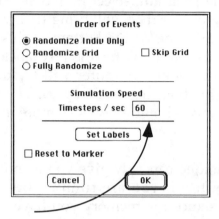

• Select the OTHER... item from the Setup menu to bring up the Other Setup box.

• Near the bottom of this dialog box is a text field labeled Timesteps/ sec. This field gives the number of timesteps EcoBeaker will try to finish each second. Type in the maximum number of timesteps per second that you want EcoBeaker to do.

• Click on the OK button.

The fastest you can get EcoBeaker to go is 60 timesteps per second, but I have yet to see a personal computer that can run even my simplest models that quickly. For most simple models, though, 2–4 timesteps per second is noticeably slower than normal, but still fast enough to be interesting. Of course, the number you type in here is a maximum speed. If you have a complicated model, then EcoBeaker will run slowly no matter how many timesteps/second you ask it to do.

Memory Matters

I rarely run out of computer memory when working with EcoBeaker, but if you create big, complicated models, then you may max out the memory that Eco-Beaker has allocated to it. You may run into problems if you have a really large grid, or if you have lots of Individualistic creatures running around in your model (thousands and thousands), or if you have many line or X-Y graphs. In any of these cases, EcoBeaker should warn you when you've run into a memory problem, and then refuse to perform the action that gave you the problem. At this point, you can either try to reduce the amount of memory you are using with your model, or you can increase the amount of memory allocated to EcoBeaker.

To increase the memory allocated to EcoBeaker, you must first quit the program. Then find the icon for the program, select it, and select the GET INFO command from the File menu of the Finder. This will bring up a window with information about your version of EcoBeaker. At the bottom of this window is an area concerned with memory allocation. Look for the box labeled Preferred size. This is the amount of memory that your computer will try to allocate for EcoBeaker. Increase the number in this box (try doubling it). Then close the info window, and run EcoBeaker again. You should now have more memory available for your models.

Note that the above technique can only give you as much memory as you have available in your computer; if you are running other programs along with EcoBeaker this will also reduce the memory you have available. Consult a reference book on the Macintosh computer for more information on memory allocation.

How EcoBeaker Works

This section gives a detailed description of how EcoBeaker constructs and runs models. It is intended for those people who may want to design their own models within EcoBeaker, either by using the built-in procedures or by programming their own procedures. It's also good reading for anyone else who will make extensive use of EcoBeaker and wants to know what is going on.

The foundation of EcoBeaker's models are the Species and the Species Grid (hereafter called the Grid, although there is also a Habitat Grid which will be mentioned later on). The Grid is the world upon which the modeling takes place. As its name implies, it is a two-dimensional rectangular surface, divided into squares. The Grid is an odd little world in which time does not advance continuously, but rather jumps from one time to the next. Each jump is called a timestep. The Species are the types of creatures that populate this world. Each square can hold one creature of a certain Species (or more than one if they are Individualistic—see below). Each Species has a set of rules associated with it which governs the actions of members of that Species.

There are two basic types of species—Grid-Based and Individualistic. The distinction between them is the way members of that species are stored within EcoBeaker. For Grid-Based species, each member of the species is stored only on the Grid, in the square that it occupies. If something else moves into that square, the Grid-Based creature that was there previously will be removed. Also, you cannot store any extra information about Grid-based creatures. However, you can store extra information for Individualistic species. With Individualistic species, each species has an array that holds the positions of all members of that species that are currently on the Grid. The advantage of this array is that additional parameters can be stored for each individual as well. For instance, you might want each individual to have a certain energy level that changes as it moves around, eats food, reproduces, or whatever. You can only do this with Individualistic species, not Grid-Based species.

Both Grid-Based and Individualistic species have three functions associated with them. The first function is performed by a Settlement procedure. The Settlement procedure is a procedure that is called at the beginning of each timestep; it can add creatures of this species to the grid. An example of a Settlement procedure would be Fixed settlement, in which a fixed number of new creatures are put in random places on the grid each timestep.

The next function is performed by the Transition Matrix. This matrix is a set of probabilities that determines whether a given creature will remain as the same species or change to a new species. If the probability of species X remaining species X is 1, nothing will happen. On the other hand, if the probability of species X changing to species Y is 30%, and the probability of changing to species Z is 20%, then at each timestep, each creature of species X will have a 30% chance of becoming species Y, a 20% chance of becoming species Z, and only a 50% chance of remaining species X. This could be used, for example, in an age-structured model, or in a model of succession.

The final function for each species is an Action procedure. This is a procedure that is invoked for each creature of that species on the Grid, and that can act on that creature in pretty much any way—make it move, reproduce, die, eat chocolate-chip cookies. The set of available action procedures is different for Grid-Based and for Individualistic species. In general, you will use action procedures only for Individualistic species.

These three functions are invoked in the following order as illustrated in the flowchart. First, the Settlement procedure for each Species is called, and any new creatures are settled onto the Grid. The different species are settled in a random order each timestep, so that no one species is always settling first; rather, each species is settled all at once. In other words, it never happens that in one timestep a creature from species 1 settles, then one from species 2, and then another from species 1.

At this point, what happens next depends on the type of randomization you have selected. With the minimum amount of randomization, each square on the Grid is examined, beginning with the upper-left corner, continuing down the leftmost column and going column by column to the right. In each square containing a creature from a Grid-Based species, first the transition matrix, and then its action procedure, is called. If the transition matrix changed the species this individual belongs to, and the new species is also a Grid-Based species, then the new species' action procedure is called.

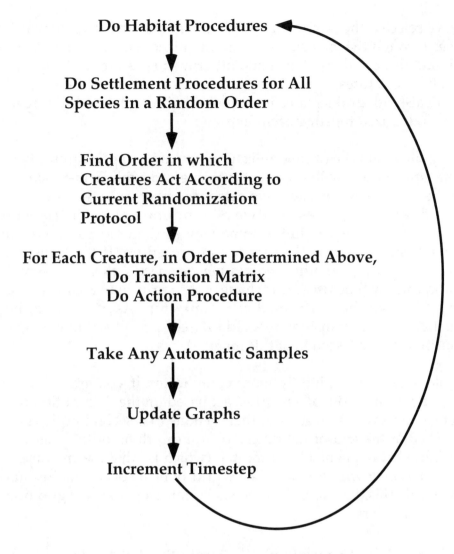

A Flowchart for EcoBeaker Models

Note that a top-left to bottom-right order of execution for Grid-Based species can create problems if the creatures on different squares interact with each other. For instance, if there is a free square with a creature on the square to the right of the free square and another on the square to the left of the free square, and both creatures want to go into the free square, the one on the left will always get to it first. Any type of movement will run into similar problems. In addition, any time there is movement to the right, you have to do some fancy programming to make sure that the creature doesn't get moved a second time, since the program is scanning the squares from left to right. Just a few warnings based on my experience.

If you have selected the Randomize Grid option, then the order in which the program goes through the grid squares is randomized each timestep, but each grid square is still dealt with as above, and all grid squares are dealt with before any Individualistic creatures. If you have selected Complete Randomization, then the program deals with grid squares interspersed with Individualistic creatures, and these are randomized together each time step.

Depending on what type of randomization you are using, creatures belonging to Individualistic species will either act interspersed with Grid-Based species, or after all the Grid-Based species have acted. In either case, the order of action of all Individualistic species is always randomized each timestep. Each Individualistic creature acts in the same way as the Grid-Based creatures, first doing its transition matrix, then its action procedure. If the transition matrix changes the creature to another Individualistic species, then the second species' action procedure will be used. If it changes the creature to a Grid-Based species, then no action will be performed that time step. As a warning, having an Individualistic species transition to a Grid Based species is a little dangerous (see the "Transition Matrix" section of this manual).

This whole algorithm is slightly more complicated if you are using a Flipping Grid. With a flipping grid, all creatures act based on the distribution of creatures in one grid, and save the results of their actions to a second grid. The grids are then switched at the end of the timestep. You can think of these as an old grid, representing what happened last timestep (which is what the creatures "look" at to decide what they will do), and a new grid (which collects the results of what the creatures do this timestep). Then the next time step, the new grid becomes the old grid and vice-versa.

Automatic sampling occurs after all the creatures have acted, and updating the graphs occurs last of all.

Included Procedures

Types of Procedures

This section of the manual lists the procedures currently included in EcoBeaker, and what each one does. There are six types of procedures:

- Settlement
- Action
- Individualistic
- Habitat
- Statistics
- Sampling

Each of these procedures will be discussed in this section in detail. You can use these procedures to make your own models without doing any programming. Note that I have NOT done extensive error-checking on most of these procedures, and that, in any case, these types of models are inherently prone to suble errors. So, if you are planning to make models using these procedures for research purposes, I suggest you run some tests to make sure each procedure does exactly what it should do. Where subtle errors are not important, however, all the procedures should work fine.

Each procedure has one or more parameters that control its operation. These parameters can be one of four types: (1) number, (2) single species or habitat, (3) list of species or habitats, or (4) an individual parameter of a species. As previously explained, when you select a procedure (or statistic or sampling technique), you will be presented with a dialog box which allows you to change feach of the parameters for that procedure. To follow is a brief description of each type of parameter, how it will appear in the dialog box, and how you change it.

Types of Parameters

Numbers

A number parameter will have a label and a text field where you can type in the value that you want. Any parameter that has a text field is a number parameter.

Single Species or Habitats

A parameter that needs a single species or habitat will be a pop-up menu. Click and hold down the mouse button on the name of the currently selected species or habitat, and a menu will pop up with a list of all the species or habitats in the model. Move the mouse to select the one you want, then let go of the mouse button.

Lists of Species or Habitats

A parameter that needs a list of species or a list of habitats will be a labeled button. Click on the button and a new dialog box will appear, with a list of each species or habitat in the model and a check-box next to each one. Check the boxes next to the species or habitats you want, and uncheck the boxes next to the ones you don't want. Then click on the OK button.

Individual Parameters

A parameter that asks for an individual param of some species will be a pop-up menu. (The individual param is a number carried by individuals of a given species.) This pop-up menu will be hooked to another parameter in the dialog box where you select the species. First select the species whose individual param you want, then click on the individual param and hold down the mouse button, so that the menu pops up. This menu will contain a list of all the individual params of this species. Move the mouse over the one you want, then let go of the mouse button to select it.

Settlement Procedures

Density Dependent

This procedure settles a fixed number of creatures per timestep, and adds to that a number of settlers proportional to the current population size of the species. For instance, you could use this procedure to simulate settlement of a plant that has a certain number of seeds blow in from the outside every timestep, but each plant that successfully settles also starts producing a certain number of seeds that land elsewhere within the world. The total number of settlers is determined by NUM IMMIGRANTS/TURN, plus the number of individuals of this species already in the world times Num Settlers/Individual.

Num Immigrants/Turn
The number of new immigrants to settle each timestep. This number of new creatures is settled into the world regardless of how many creatures of this species are already in the world. If this number is less then 1, then it represents the probability per timestep that a single creature of this species will settle.

Num Settlers/Individual
The number of new creatures added to the grid every timestep, per individual that already exists. You can think of this as an intrinsic growth rate for the species (assuming there are no other ways that the species can reproduce, such as through an action procedure)

Possible Uses
I use this procedure to make plants whose growth depends on their density in the model, and who have no action procedure. When I make animals, they usually have density-dependent growth by virtue of their action procedure, so I usually don't use this settlement procedure for species that have an action procedure.

Fire

This procedure makes fires that burn everything in their path. The chance of a fire starting each timestep is given in Chance of Fire, and the chance of it spreading from each square it reaches to neighboring squares is given in Chance of Spread. There is no way to make one creature more fire-resistant than others. However, if there are two or more individualistic creatures in one square, only the top one will get burned. The others are hidden from the fire underneath the first one.

Chance of Fire
The probability that a fire will start somewhere in the grid, per timestep.

Chance of Spread
Once a fire starts, the chance that it will spread to each neighboring square. From each square that a fire reaches, it can spread to any or all of its four nearest neighboring squares, with the chance of spreading to each possible neighbor independent of the others.

Note: You'll notice a "percolation" effect here. There is a threshold (around 0.75) above which fires spread much farther than they do with lower values.

Possible Uses
In order to make fires, I make a Fire species with this settlement procedure whose transition matrix specifies that a Fire square immediately turns into an Empty square (transition to Empty is 1.0). This makes a fire that wipes out everything in its path and leaves empty ground behind. You could also use this procedure in other places where you wanted a species to quickly spread out over an area, or for other types of local disturbances.

Fixed

This is the simplest settlement procedure in the program. It simulates immigration from outside the modeling world at a constant rate per timestep. Num Immigrants/Turn sets the number of new settlers coming into the world per timestep, and Can Settle Over sets the list of species that the immigrants can successfully settle over. A random square is picked for each settler, and if that square is already occupied by a creature that is not in the Can Settle Over list, that settler won't be able to settle. Otherwise, it kills the creature already at that square (if the creature is Grid-Based) and settles over it.

Num Immigrants/Turn
The number of new creatures to settle each timestep. If this is less then 1, this number represents the probability per timestep that a single creature of this species will settle.

Can Settle Over
A list of the species that these settlers can successfully settle on top of. If a settling creature lands on top of a species that is not on this list, then the settler will not survive.

Possible Uses
Use this procedure for any species for which you want a constant immigration rate. You might use it, for instance, for a plant species that has a fixed number of seeds blowing in and settling every timestep. You might also use it for an animal that has a low probability per timestep of having a new colonist arrive.

Fixed @ Time X

This procedure performs a single bout of settlement of a fixed number of creatures. Num Immigrants creatures are settled onto random places in the grid at timestep Timestep to Settle. A creature can only settle successfully if it lands on a square containing a species in the Can Settle Over species list. If one of the settlers lands on top of a species not in this list, that settler dies.

Num Immigrants
The number of creatures of this species to settle. If this is less than 1, this number represents the probability that a single creature of this species will settle.

Timestep to Settle
The timestep at which to settle the creatures. Settlement happens only once.

Can Settle Over
A list of the species that these settlers can successfully settle on top of. If a settling creature lands on top of a species that is not on this list, then the settler will not survive.

Possible Uses
I use this procedure a lot for adding a given number of creatures into a model at the start of a run when I don't want constant immigration. Sometimes I delay the settlement until some other event happens. For instance, in the case in which one species eats another species, I may want to give the prey species time to build up its population size.

Hab Density Depend

This procedure is like Density Dependent settlement, with the added feature that creatures have varying chances of settling, depending on the type of habitat they land in. Num Immigrants/Turn gives the number of new creatures that will land in the Grid each timestep. Or if it's less than 1, then this number represents the chance that a creature will immigrate into the world per timestep. Added to this basal rate of settling is an additional number—Num Settlers/Individual—multiplied by the population size of this species. Each new creature settles in a random place in the Grid, and the chance of it successfully settling is given by Chance of Settling for each habitat type.

Num Immigrants/Turn

The number of new immigrants to settle each timestep. This number of new creatures is settled into the world regardless of how many creatures of this species are already in the world. If this number is less then 1, this number represents the probability per timestep that a single creature of this species will settle.

Num Settlers/Individual

The number of new creatures added to the grid every timestep per individual that already exists. You can think of this as an intrinsic growth rate for the species (assuming there are no other ways that the species can reproduce, such as through an action procedure).

Chances of Settling

A list of the probabilities that a creature landing in each habitat type will successfully settle. The chance can range from 0 (never successful) to 1 (always successful).

Possible Uses

This is a rather special-purpose settlement procedure, useful when you want both density dependence and certain habitats to be better than others.

Habitat Fixed

This procedure is exactly like the Fixed settlement procedure, except that the creature has different chances of settling, depending on what type of habitat it lands on. Num Immigrants/Turn tells the number of creatures of this species that will be dropped in random places on the Grid every timestep. The chance of a creature successfully settling on each habitat type is given in Chances of Settling.

Num Immigrants/Turn

The number of new creatures to settle each timestep. If this is less then 1, this number represents the probability per timestep that a single creature of this species will settle.

Chances of Settling

A list of the probabilities that a creature landing in each habitat type will successfully settle. The chance can range from 0 (never successful) to 1 (always successful).

Possible Uses

I use this procedure in several models to settle food on certain habitats and not others—for instance, on islands in an ocean or on fertile ground in the midst of not-so-fertile ground. One nice feature of the procedure is that the settlement rate per unit area will remain constant as you change the areas of the good and bad habitats (because a constant number of immigrants are raining down over the whole grid).

Hab Fixed @ Time X

This procedure is exactly like Fixed @ Time X settlement procedure, except that the chance of the creature successfully settling depends on what habitat type it lands on. Num Immigrants gives the number of creatures of this species that will be settled at the timestep given in Timestep to Settle. The chance of a creature successfully settling on each habitat type is given in Chances of Settling, where 0 = no chance of settling if it lands in that habitat, and 1 = always successfully settles in that habitat.

Num Immigrants
The number of new creatures to settle each timestep. If this is less then 1, this number represents the probability per timestep that a single creature of this species will settle.

Timestep to Settle
The timestep at which to settle the creatures. Settlement happens only once.

Chances of Settling
A list of the probabilities that a creature landing in each habitat type will successfully settle. The chance can range from 0 (never successful) to 1 (always successful).

Possible Uses
Use this procedure when you want a combination of a single bout of settlement settlement with different settling chances on different habitats.

Habitat Timed

This settlement procedure settles creatures periodically while the model is running. Num Immigrants/Turn tells the number of individuals that are settled each settlement event. The first settlement happens at timestep Time of First Settling. From then on, settlement will occur at intervals of Time Between Settling timesteps for as long as the model is run. For instance, if Time Between Settling is set to 10, then creatures will be settled at timestep 0, timestep 10, timestep 20, and so on. The chance of a creature successfully settling is dependent on the type of habitat it lands in, as given in Chances of Settling.

Num Immigrants/Turn
The number of creatures of this species that will be settled onto the Grid during each settling event. Each individual will be dropped onto a random place onto the Grid. If this number is less than 1, it represents the chance that a single individual will be settled at each settling event.

Time of First Settling
The timestep of the first settling event.

Time Between Settling
The number of timesteps between settling events. The first settling event will occur at Time of First Settling. The second event will occur at the time given by Time of First Settling plus Time Between Settling, and so on.

Chances of Settling
A list of the probabilities that a creature landing in each habitat type will successfully settle. The chance can range from 0 (never successful) to 1 (always successful).

Possible Uses
I use this procedure when I want to have some settlement event occur on a regular basis, but not at every timestep. For instance, if I want settlement to occur once a day and the timestep is 1 hour, then I would use this settlement procedure with Time Between Settling equal to 24, or if I wanted settlement once a year and the timestep was 1 month, I would use a Time Between Settling of 12. Many times I don't care about the habitat aspect of this settling procedure, so I make Chances To Settle equal to 1 for all habitats.

Hurricane

This procedure makes hurricanes that sweep through the world and kill a percentage of creatures from any affected species. The chance of a hurricane occurring in a timestep is given by Chance of Hurricane, the chance that a creature will die is given by Chance of Death, and the species affected are given in Susceptible Species. When there is a hurricane, it visits each square on the Grid. You'll see this as a flashing of each square into the color of the Hurricane species. If the creature in that square is a member of a susceptible species, then it has a chance of dying, given by Chance of Death. If there are two or more individualistic creatures in a single square, only the top one will be affected by the hurricane; the others are "hidden" under the top one.

Chance of Hurricane
The chance in each timestep that a hurricane will occur.

Chance of Death
The chance that a creature will die during a hurricane.

Susceptible Species
A list of the creatures susceptible to hurricanes.

Possible Uses
This procedure can be used to simulate any general density-independent death process.

Interact

This procedure makes the chance of settling successfully a function of how close your nearest already-established neighbor is. Every timestep, a number of potential settlers given by Num Immigrants/Turn immigrate into the model, landing in random locations on the Grid. For each settler, the procedure looks to find the nearest neighbor within the number of squares given by Max Effect Distance. If there are no neighbors within this area, then the nearest neighbor is assumed to be at the Max Effect Distance. The chance of settling successfully if the nearest neighbor is 0 squares away (if you landed on top of an already-established creature) is given in Chance @ 0 Dist, and the chance of settling successfully if the nearest neighbor is the Max Effect Distance away is given in Chance @ Max Dist. The chance of settling with a nearest neighbor somewhere between these two distances is a linear interpolation between the two chances. By setting the chances one way, you'll get facilitation, the other way you'll get inhibition of nearby settlement. The Affected By button lets you set which species will affect the settlement chance of this species, and the Dominant Over button lets you set which species a new settler of this species can displace.

Num Immigrants/Turn
The number of creatures in this species that will be settled onto the Grid during each settling event. Each individual will be dropped randomly onto the Grid. If this number is less than 1, then this number represents the chance that a single individual will be settled at each settling event.

Max Effect Distance
The maximum number of squares away an already-established creature can affect a new immigrants chances of settling.

Chance @ 0 Distance

The chance that an immigrant will settle successfully if it is landing on top of a creature that can affect it.

Chance @ Max Distance

The chance that an immigrant will settle successfully if the nearest creature that can affect it is Max Effect Distance squares or more away from it. For distances between 0 and Max Effect Distance, the chance of settling successfully is a linear interpolation between Chance @ 0 Distance and Chance @ Max Distance.

Affected By

A list of the species that can affect the successful settling of immigrants of this species.

Dominant Over

A list of the species that creatures of this species can settle successfully on top of. If a settling creature lands on top of a species that's not on this list, it will not survive.

Possible Uses

I use this procedure primarily to make some species settle in a patchy or even distribution. You can also get interesting interspecific patterns by making two species reciprocally Affected By each other.

Random Interact

This procedure is very much like the Interact procedure described above. The only difference is that, instead of a fixed number of individuals being settled each timestep, the number of settlers per timestep is picked from a distribution of possible numbers. Since the explanation of Interact settlement is a bit long, I'll refer you to the Interact procedure above and only explain the differences here. The number of settlers per timestep is a random number between Min Num Immig/Turn and Max Num Immig/Turn. If Num Random Picks is more than 1, then the algorithm picks Num Random Picks random numbers between the min and max numbers, and averages all these random numbers to determine the number of settlers. This calculation creates an approximation to a normal distribution, with higher Num Random Picks distributions having more chance of getting the average number of settlers.

Min Num Immig / Turn

The minimum number of new settlers in each timestep.

Max Num Immig / Turn
The maximum number of new settlers in each timestep.

Num Random Picks
The number of times to pick a random number between MIN NUM IMMIG/TURN and MAX NUM IMMIG/TURN. The average of NUM RANDOM PICKS random numbers is the number of creatures that will settle in a given timestep. As NUM RANDOM PICKS goes up, the variance around the average goes down and there is more chance of the number of settlers being halfway between the min and max.

Other Parameters
See the "Interact Settlement" procedure above.

Possible Uses
Use this when you want an interact-type settlement procedure (perhaps to make a nonrandom spatial distribution of creatures), but also want the number of settlers in each timestep to be variable.

Random Number

This procedure settles a random number of individuals onto the Grid each timestep. Each individual settles at a random place on the Grid. The number settled per timestep ranges between Min Num Immig/Turn and Max Num Immig/Turn. The mean number of individuals settled per timestep will be halfway between the minimum and maximum. You can specify the variance around this mean by using the Num Random Picks parameter. If Num Random Picks is 1, the procedure will pick a single random number between Min Num Immig/Turn and Max Num Immig/Turn, and settle that many individuals. If Num Random Picks is 2, the procedure will pick two random numbers between the minimum and maximum, average them, and settle that many individuals. This averaging has the effect of reducing the variance. The higher you set Num Random Picks, the lower will be the variance around the mean number of settlers.

Min Num Immig / Turn
The minimum number of new settlers in each timestep.

Max Num Immig / Turn
The maximum number of new settlers in each timestep.

Num Random Picks
The number of times to pick a random number between Min Num Immig/Turn and Max Num Immig/Turn. The average of Num Random Picks random num-

bers is the number of creatures that will settle that timestep. As Num Random Picks goes up, the variance around the average goes down and there is more chance of the number of settlers being halfway between the min and max.

Can Settle Over
A list of the species that these settlers can successfully settle on top of. If a settling creature lands on top of a species that is not on this list, the settler will not survive.

Possible Uses
Use this procedure when you want a species whose immigration varies from timestep to timestep.

Restricted

This settlement procedure drops a number of new creatures given by Num Immigrants/Turn onto the Grid each timestep. The creatures are dropped in random locations within the rectangular region specified by Minimum X, Maximum X, Minimum Y, and Maximum Y. They only settle successfully if they land on top of a species that is listed in Can Settle Over.

Num Immigrants/Turn
The number of new creatures to settle each timestep. If this is less then 1, then this number represents the probability per timestep that a single creature of this species will settle.

Can Settle Over
A list of the species that creatures of this species can successfully settle on top of. If a settling creature lands on top of a species that is not on this list, the settler will not survive.

Minimum X
The left side of the area where settlers will be dropped.

Maximum X
The right side of the area where settlers will be dropped.

Minimum Y
The top of the area where settlers will be dropped.

Maximum Y
The bottom of the area where settlers will be dropped.

Possible Uses

I use this instead of the Fixed settlement procedure when I want a fixed number of immigrants per timestep, but only want them in a specific part of the Grid. For instance, you could use this to make a patch of food, or to have immigration occur only through a certain area of the Grid.

Tide

This settlement procedure is supposed to simulate a tidal cycle. Every timestep, air will come down from the top of the Grid and then disappear again, simulating the tide going out and then coming back in. At low tide, the air will come down the number of squares given by Min Tide Height. At high tide, the air will come down Max Tide Height squares. The length of one cycle from low to high and back to low tide is Tidal Cycle Length timesteps. Each time the air hits a creature, that creature has a chance of dying, as given in Chance of Death.

Min Tide Height

The number of squares down from the top of the Grid that the tide goes at its lowest.

Max Tide Height

The number of squares down from the top of the Grid that the tide goes at its highest.

Tidal Cycle Length

The number of timesteps it takes for the tide to go from its minimum height to its maximum height and back.

Chance of Death

A list giving the probability of death for creatures of each species when a tide hits them.

Possible Uses

I've only used this once, to make a tide with a 28-day cycle that killed some species more than others.

Grid-Based Action Procedures

There are two types of Action procedures to go along with the two types of Species—Grid-Based and Individualistic. The Action procedures for Grid-Based species are not very useful, and there are currently only four.

Cell Automaton

This procedure implements the Game of Life (an old AI game that has very little to do with real life and a lot to do with too many candy bars late at night). It's a very basic cellular automaton procedure that looks at the eight neighboring squares and determines whether the square in the middle will be species A or species B based on how many of the neighbors are species A.

This procedure looks at the eight squares surrounding each square, and counts the number of creatures in those eight squares that belong to one of the species listed in Affectors. If that number is more than or equal to the number given by Min Creatures and less than or equal to Max Creatures, the central square will become a Plus Species; otherwise, the central square will become a Minus Species.

Min Creatures
The minimum number of creatures that must be surrounding this square for it to be a member of the Plus Species next timestep.

Max Creatures
The maximum number of creatures that can surround this square for it to be a member of the Plus Species next timestep.

Plus Species
The species that this square becomes if it is surrounded by a number of creatures somewhere between Min Creatures and Max Creatures.

Minus Species
The species that this square becomes if it is surrounded by a number of creatures less than Min Creatures or more than Max Creatures creatures.

Affectors
A list of the species that are counted in determining the state of this square next timestep. Usually this will be the Plus Species.

Possible Uses
See my Game of Life situation file (it's on my web site) for an example of how to use this procedure. Basically, you will want to make two species, and give them both this procedure with exactly the same parameters in each species. Then you can paint a few individuals of the Plus Species onto the grid, and let it run. You should probably be using a flipping grid with this procedure.

Note: There are lots of programs available specifically to run the Game of Life, and to do other cellular automata. EcoBeaker is not optimized to do cellular automata models, so if you want to do a lot of these you might look for another program.

Interact Death

This procedure kills creatures based on how close they are to other creatures. Depending on how you set the parameters, you can make the chance of death go up or down as the distance to the nearest neighbor decreases. This procedure is kind of an analog of the Interact Settlement procedure, except that it kills adults instead of settlers. See the "Interact Settlement" procedure for a further explanation.

Max Effect Distance
The maximum number of squares away that another creature can be and still affect the chance of death of this creature.

Chance @ 0 Dist
The chance of death of this creature when it is on top of another creature that can affect it.

Chance @ Max Dist
The chance of death of this creature when the nearest other creature that can affect it is Max Effect Distance squares or more away from it. For distances between 0 and Max Effect Distance, the chance of death is a linear interpolation between Chance @ 0 Dist and Chance @ Max Dist.

Affected By
A list of the species that can affect the death of creatures of this species.

Possible Uses
You can use this to make the rate of death of a species affected by neighboring individuals of the same or other species. This can give different types of patchy or even distributions of creatures. I've also used this together with the Interact Settlement procedure to have both settlement and death influenced by local

surroundings. Be a little careful with this procedure—if you don't randomize the grid, the action procedures occur in a wave going across the screen from left to right, so the creatures on the left are more vulnerable to death from this procedure than those on the right.

Mutate

This procedure gives a chance that a creature of this species will change into a creature of another species. It's like using the transition matrix, and is useful only if you might want two transitions in a single timestep. Set Mutate Chance to the chance that a creature will become another species, and Mutate To to the species that the creature should change into.

Mutate Chance
The chance per timestep that a creature of this species will change into a creature of the Mutate To species. This is exactly the same as setting the appropriate probability in the transition matrix.

Mutate To
The species into which creatures of this species will mutate into.

Possible Uses
This procedure is a bit of a relic, and I almost never use it.

Die At

This procedure will kill all individuals of this species at periodic intervals. Every time the number of timesteps given in Die Times is reached, all creatures of this species will be killed.

Die Times
The time from one die-off to the next. All creatures of this species will be killed at timestep Die Times, at the timestep equal to twice Die Times, three times Die Times, and so on.

Possible Uses
I've used this procedure as a rudimentary way to automatically run a model several times in a row. For instance, if I want a model to run for 1000 timesteps, output a result, then run again, I set DIE TIMES equal to 1000 for some species whose disappearance will reset the model, make a graph that only prints out results at intervals of 1000 timesteps. Although this is clumsy, it has worked for me. In the next version of EcoBeaker I hope to include a far less clumsy macro language.

Individualistic Action Procedures

These are only used for Individualistic species.

Many of these procedures are offshoots of the Predator procedure, so it might be worthwhile to read the Predator procedure first. To avoid repeating the same information, I frequently refer to the Predator procedure in this section.

Commonly Used Individual Parameters

Each of these procedures has a number of individual parameters. Individual parameters are numbers that are stored separately for each individual of the species. For instance, each individual stores its own *x,y* position in the grid. You can see these individual parameters for a given creature by clicking on it on the Grid (Getting Info About Creatures), or there are some statistics that will let you look at the average value or the distribution of an individual parameter. I list here some of the individual parameters that are used in many or all of the procedures below. For other individual parameters that are used only in one action procedure, I give descriptions within the text describing that action procedure.

X Pos
> The *x* position of this individual on the grid.

Y Pos
> The *y* position of this individual on the grid.

Energy
> The energy level of this individual. In most procedures that use energy, energy is gained by eating, a little is lost every timestep for maintenance of the individual, and the rest is used for reproduction. You could think of this as the amount of fat the creature has.

Temp Param
> This is my name for a parameter that is needed internally by the procedure, but isn't interesting in its own right. You can ignore parameters with this name.

Biased Predator

Biased Predator is very similar to the Predator procedure; refer to the Predator procedure for a further description. The difference between the two is in what creatures do when they don't see any food. When you use Predator procedure, a creature that doesn't see any prey items will do a random walk. However, when you use Biased Predator procedure, the creature does a biased random walk. Thus, if Move Randomness is set to 0, the creature will randomly pick a direction and speed, and walk in that direction until it sees a prey item. If Move Randomness is set to 1, the creature will again pick a random direction and speed, but now the direction and/or speed will be modified by up to one square each timestep. Higher numbers for Move Randomness mean less bias to the direction of the random walk.

Move Randomness
Each individual has a movement vector that determines the number of squares it moves each timestep to the right or left, and the number of squares it moves up or down. Every timestep, a number of squares given by Move Randomness is added or subtracted to each component of the movement vector.

Other Parameters
See the parameters of the same names in the Predator procedure.

Special Individual Parameters
Each individual stores a Move Vector X and Move Vector Y, which together indicate the direction in which the individual is currently moving.

Possible Uses
Use this procedure in place of the Predator procedure when you want creatures to move in straighter lines while looking for food (straighter than they would if you had selected the Predator procedure.)

Disperser

This procedure is supposed to model species when we are interested in studying dispersal behavior through a fragmented landscape. Individuals of this species have a Death Rate and a Chance of Movement, which are based on the type of habitat they currently occupy. If an individual moves, which occurs based on a probability given in Chance of Movement, it moves a number of squares (given by Speed) in a random direction. Movement itself contributes to the chance of this individual's death. If an individual moves, its chance of death increases up by Movement Death Rate.

Individuals can also be territorial, as specified in Interact Dist and Max Interact Death. If an individual is within Interact Dist of another individual of the same species, it will try to move away. Its chance of death also increases based on how close it is to another individual of the same species. If two individuals occupy the same square, each of their chances of death increase by Max Interact Death. If two individuals are in different squares but still within Interact Dist of each other, their death rates are higher by a fraction of Max Interact Death, proportional to how far away from each other they are (going to 0 as their separation approaches Interact Dist).

A species with this action procedure reproduces every Birth Timesteps, and each time it reproduces it has Num Offspring/Time babies. For instance, if Birth Timesteps is equal to 10 and Num Offspring/Time equals 4, then each individual alive at timestep 0 will have four babies, individuals alive at time 10 will each have four babies, individuals alive at time 20 will each have four babies, and so on.

Chance of Death
A habitat-specific list of the chance per timestep that creatures of this type will die. Each habitat can have a different chance of death.

Chance of Movement
A habitat-specific list of the chance per timestep that creatures of this type will move. Each habitat can have a different chance of movement. These chances are multiplied times the overall Max Chance Move parameter.

Movement Death Rate
The increase in the chance of death when a creature moves. This is added to the Chance of Death.

Interact Distance
The number of squares apart two creatures of this species can be and still interact.

Max Interact Death
The maximum increase in the chance of death when two creatures of this species interact. This maximum occurs when the two creatures are on top of each other. As the distance between a creature and its nearest neighbor increases, the additional chance of death due to interaction decreases. The chance of death due to interaction goes to 0 when the nearest neighbor of a creature is Interact Distance squares away.

Max Chance Move
The maximum chance that a creature will move. This chance is modified by habitat-specific probabilities, given in Chance of Movement.

Speed
The number of squares per timestep that a creature of this type can move.

Num Offspring/Time
The number of offspring a creature of this type has each time it reproduces. These offspring are deposited on top of the parent creature.

Birth Timesteps
The number of timesteps between reproduction events. All creatures reproduce synchronously at the timestep equal to Birth Timesteps, the timestep equal to twice Birth Timesteps, and so on.

Species of Babies
The species of the babies made by creatures of this type. The species can be the same as or different from the parent species.

Possible Uses
I use this procedure for a territorial species in a fragmented habitat. You can easily take out certain aspects of the procedure by setting some parameters to 0. For instance, set Interact Distance to 0 to make nonterritorial creatures; set Movement Death Rate to 0 if you don't want movement to contribute to death.

Evolve Movement

This special-purpose procedure is useful when you wish to look at the evolution of search behavior. It gives each individual creature a search strategy to use in looking for food; this search strategy gets mutated slightly in each of the creature's offspring, which means that the search strategy can evolve over time. The search strategy is very simple. A creature can move in one of six directions each timestep, relative to the direction that it moved the last timestep. It can keep going in the same direction (forward); it can turn around 180 degrees and go back to where it came from (backward); it can turn slightly to the right or left (fore-right and fore-left); or it can turn considerably to the right or left (back-right and back-left). Whenever it moves on top of one of the species given in Prey Species, it eats Prey Species and gains Prey Value energy units. Each timestep the creature uses up Cost of Living units of energy to stay alive; if a creature's energy level falls below 0, it dies. If a creature's energy level rises to Energy to Reproduce, the creature splits in half, and gives half of its energy to each of its daughters.

Each creature of this type has a set of six probabilities giving the chance that it will go in one of six directions. This set of probabilities, called individual parameters, is different for each individual. The set of probabilities adds up to 1, which means that the creature will move one square in a given direction each timestep. Each time it reproduces, a creature passes down its set of probabilities to its baby, with a slight mutation. The mutation reduces the probability that it will go in one of the directions, and increases the probability that it will go in another one of the directions by that same amount (to keep the total equal to 1). The maximum size of the mutation is given by % Mutation/Generation, and the directions which are mutated are selected randomly. Initially, creatures start out with approximately equal probabilities of going in any direction.

Cost of Living
The number of energy units a creature spends each timestep in order to stay alive.

Prey Value
The number of energy units gained from each prey item eaten.

Energy to Reproduce
The amount of energy these creatures need to reproduce. When the energy level of a creature reaches this level, it reproduces by splitting in half, with each daughter getting half the energy. New immigrants (which came in with a settlement procedure) start out with an energy level equal to one-half of Energy to Reproduce.

% Mutation / Generation
The maximum mutation that can occur each time a creature reproduces. For instance, if this total is 1, then up to a 1% chance can be subtracted from one of the directions and added to another one.

Prey Species
The species (only one) that creatures of this type will eat.

Special Individual Parameters
Each individual stores the Direction in which it is moving, which is a number from 1 to 6 (where 1 means going down the screen, 4 means going up the screen, and the remaining numbers are filled in clockwise).

Possible Uses
This is a very special-purpose procedure which shows the action of natural selection on a simple search strategy. You can alter it a little by having the creatures evolving in different types of environments—random distributions of food versus patchy distributions or only one patch, or a particularly good patch of food within an area of sparser food. I got the idea for this procedure from an old *Scientific American* article on evolving "bugs" in a computer.

Evolve Repro Energy

This procedure is very similar to the Predator action procedure, with the added feature that the amount of energy a creature needs to accumulate before reproducing is an individual parameter and can evolve. Creatures of this type look for food. They can see Distance to Look squares in all directions. If they see food (as defined in Prey Species), they move towards it at the number of squares per timestep given in Speed. When they eat an item of food, they gain Prey Value energy units, and each timestep they use Cost of Living energy units to stay alive. If they reach their reproductive energy threshold, they reproduce, and give half of their energy to each daughter. Each individual has its own reproductive energy threshold; the daughters inherit this threshold from the parent, with a slight mutation. This mutation means that the reproductive energy threshold evolves over generations.

Initial Reproduce Energy
The reproductive energy threshold originally given to creatures of this type when they settle onto the Grid (with a settlement procedure).

Mutate Repro Energy by
The amount by which the reproductive energy threshold is mutated from parent to daughter. This parameter gives the maximum amount of mutation; the mutation can be either positive or negative.

Other Parameters
See the parameters of the same names in the Predator procedure.

Special Individual Parameters
Each individual stores its own Repro Energy—the amount of energy that it will build up before reproducing.

Possible Uses
I use this in a lab that looks at *r* vs. *K* selected species. I think this procedure gives a particularly clear demonstration of natural selection in action.

Evolve Sex Ratio

This special-purpose procedure is useful when you wish to look at the evolution of sex ratios. This procedure is similar in many ways to the Predator procedure, so refer to the Predator procedure for a more thorough description of some of the parameters. Every timestep, each creature of this type looks for the closest piece of food within Distance to Look squares, and moves towards that food with a speed given by Speed squares per timestep. Each timestep it loses Cost of Living energy units, and each time it lands on food (specified in Prey Species) it gains Prey Value energy units. If it falls below 0 energy units, the creature dies.

The difference between this procedure and the Predator procedure is in reproduction. Each species given this procedure is either male or female, and you must have both a male species and a female species in the model. If an individual goes above Min Repro Energy energy units, it starts looking for a mate. The mate species is specified in Mate Species, and MUST be another species in the model that also has the Evolve Sex ratio action procedure, and is of the opposite sex. If an individual reaches the same square as an individual from the Mate species, and they are both above their respective Min Repro Energy, they reproduce.

The number and sex of offspring are determined by the female. The female determines the sex of the babies based on her sex-ratio "gene." The population can have two alleles of this gene initially, which are set in Chance of Female 1 and Chance of Female 2. Both Chances are numbers between 0 and 1, which determine the chance that an offspring will be female. Any creatures introduced into the model with a settlement procedure (as opposed to actual offspring) will be randomly assigned one of these two alleles. Each offspring is randomly given the allele from either its father or its mother, so the frequency of the two alleles in the population can evolve over time.

A few more details. For each offspring produced, the mother and father each expend Energy per Offspring units of energy (which differs for the two sexes). Each new offspring starts off life with 10 units of energy (this is not a user-settable parameter).

Female ('0') or Male ('1')
Make this 0 if this is the female species or 1 if this is the male species. A model should have both a male and a female species.

Min Energy to Repro
The minimum energy that a creature needs to accumulate before it can reproduce. Once it accumulates this much energy, the creature will start looking for another creature to mate with. As it looks for mates, it also keeps looking for food. This energy threshold can be different for males and females.

Num Offspring/Mating
The number of offspring produced from a mating. If this number differs for males and females, the female number is the one used.

Energy per Offspring
The amount of energy used by this creature for each baby it makes. This can differ for males and females.

Chance of Female 1
One of the sex-ratio alleles. This is the chance of a new baby being female. 0.5 means this gene specifies equal numbers of females and males, 0.75 means the gene specifies 3 times as many females as males, etc.

Chance of Female 2
The other sex-ratio allele. See Chance of Female 1 above.

Mate With
The species of the opposite sex that this species should mate with.

Other Parameters
See the parameters of the same names in the Predator procedure.

Special Individual Parameters
Each individual stores its own male/female gene as a Chance Female parameter. It also stores the number of babies it will produce as Num Kids, the energy it will spend on each as Energy/Kid, and the last timestep in which it mated as Last Mating.

Possible Uses
I've used this to make a lab to examine why almost all species have a 50/50 sex ratio. I've also tried to use it to look at why group selection might favor a female-biased sex ratio. Both of these lab examples, especially the second, have had mixed results, primarily because evolution takes a long time, even on a computer, and because the "wrong" answer can occur quite frequently due to chance in a model that is small enough to run in a reasonable amount of time. Still, I've had fun playing with this procedure.

Note: By setting the Chance of Female parameters differently for the males and females, you can have up to four different alleles in the population instead of two. Also, you can use this as a general-purpose sexual predator procedure if you set all alleles to 0.5, so there will be no evolution (though you could just as well use the Sexual Predator procedure).

Forager

This procedure is meant to simulate a foraging organism, such as a bee, which goes in search of food, gathers up this food until a certain condition is met, then returns home and drops off its load. Each individual of this species starts out with an Energy level of 0. Each timestep, the individual looks for food in a number of squares (Distance to Look); if it sees food within that area it moves towards the food at Speed squares per timestep. You can specify which species count as food, and how many energy units each food species is worth, in the Prey Species dialog box. The numbers in Prey Species are the energy value of each species in the model, and a 0 means that species is worth no energy and thus is not food. Each timestep, the organism uses up Cost of Living units of energy to keep moving.

The home of an individual is the square where it started out. The only exception to this is if it starts out in an Empty square, in which case the individual's home is assumed to be outside the Grid. In the latter case, when the individual returns home it will be deleted from the model.

The tricky part of the Forager procedure is specifying the rules that determine when an individual will return home. There are six sets of rules; you specify which one you want to use by typing in the appropriate number for Return Home Type.

Assuming that the individual did not start in an Empty square, when it returns home it dumps whatever food it has, and its Energy level goes back to 0. Then it sets off again in search of more food.

Distance to Look
The number of squares in which these creatures will look for food, each timestep.

Speed
The maximum speed, in squares, that a creature of this type can move in one timestep.

Cost of Living
The number of energy units a creature spends each timestep in order to stay alive. Note that, unlike some other action procedures, creatures in this action procedure will not die even if they go below 0 energy level. If you are interested in how much food they collected, but not how much energy it cost to collect that food, set Cost of Living to 0.

Prey Values
A list of the number of energy units a creature of this type will get from eating each other species in the model. These creatures will only look for species whose energy value is more than 0, but they will not preferentially go towards higher-valued food items.

Return Home Type
This parameter determines when a creature will stop foraging and return to its home base. It can be set to one of the following rules (use the number next to the rule you want). Most of these rules have a parameter associated with them, which you put into Parameter 1 (and for rule number 5, also into Parameter 2).

(0) Never return home.

(1) Return home after going above some absolute energy threshold.

(2) Return home when the rise over the run of the energy level (from the time it reached the first particle of food) goes below a certain value.

(3) Return home after a certain amount of time has passed.

(4) Return home after going more than a certain number of timesteps without eating.

(5) Return home when the average feeding rate over a certain time period falls below a certain percentage of the maximum possible feeding rate. This option takes two parameters—the number of timesteps and the threshold percentage. Note that this is a running average—not a true average—this "average" declines exponentially.

Parameter 1
This is the parameter that is linked to the rule you choose in Return Home Type. For example, the threshold for rule 1, the value for rule 2, the amount of time for rule 3, the number of timesteps for rule 4, and the time period to average over for rule 5.

Parameter 2
Used only for rule 5, in which case this is the threshold percentage.

Special Individual Parameters
Each individual stores a Variable, whose meaning depends on Return Home Type, and a Mode, which tells it whether it's looking for a patch, foraging, or returning home (1, 2, or 3, respectively). Each individual also stores the position of its home base as Home X, Home Y. If these are –10000, there is no home base.

Possible Uses
This procedure has been used to look at different types of foraging strategies in different environments, and to get students to try to guess what foraging

strategy a particular creature is using by having students examine the food-collection curve and determine the Return Home Type Parameter.

Picky Predator

This procedure makes a predator whose different food items are not worth equal amounts of energy. This predator preferentially seeks higher-energy food. This procedure is similar to the Predator procedure; look at the description of Predator for parameters that are not fully explained here. Every timestep, each creature of this type looks for the closest piece of food within Distance to Look squares away, and moves towards that food with a speed of Speed squares per timestep. Each timestep it loses Cost of Living energy units to stay alive, and each time it eats some food it gains energy. The Prey Values parameter gives the energy value of the other species in the model. Only species with a prey value above 0 will be considered food.

A creature with this procedure will first look for the prey with the highest prey value. If it doesn't see one of these, it looks for the next highest value, and so on. When a creature with this procedure eats a prey item, it remembers the species of that prey. For a given number of timesteps after that, it will look only for prey items of that species. This number of timesteps is given in Time Until Switch Prey. After Time Until Switch Prey timesteps have passed without finding any prey, the creature will look for other types of prey, in order from highest to lowest prey values.

Note: If a creature with this procedure is looking for one type of prey, it will ignore other potential prey items and will even walk over them without eating them. Also, if two prey items have the same value, the creature will preferentially go for the one that is higher on the species list (nearer to the top of the Species window).

Prey Values
A list of how much energy each species in the model is worth as food to this species. Any species whose prey value is 0 will not be considered food. Picky Predators look first for the prey whose energy value is highest; if they don't see any, they look for the next highest energy value, and so on. If two species have the same energy value, the one with the lower species number (higher up in the list) will be preferred.

Time Until Switch Prey
The number of timesteps a creature will go without eating before it switches the type of prey it's looking for. If Time Until Switch Prey is –1, the creature will always look for all possible prey each timestep, seeking the one with the highest

energy value. If Time Until Switch Prey is 0, a creature will first look for the prey species it ate last timestep, but if it doesn't find it, it will immediately start looking for other types of prey. If Time Until Switch Prey is 1 or more, the creature will keep looking for the prey type of its last meal for this number of timesteps, after which it will start looking for any type of prey.

Other Parameters
See the parameters of the same names in the Predator procedure.

Special Individual Parameters
Each individual stores the amount of time that has passed since it last ate as Time Since Ate, the last species it ate as Last Spec Eaten, and the species it ate last timestep, if any, as Gut Contents.

Possible Uses
This is a slightly more sophisticated version of the Predator procedure, which takes into account that not all types of food are equal and that a creature might need to search for only one type of prey at a time. I've used it primarily to make different prey worth different amounts. I've also used it to look at optimal foraging-type problems, changing the prey switching times, the prey values, and the distribution of prey.

Predator

This procedure is supposed to simulate a creature that moves around and eats other creatures, reproduces, and dies of starvation. Each timestep, creatures using this procedure search a distance that is Distance to Look squares away, and find the nearest food item. They then move towards that food item at a speed designated by Speed. Each time they move they use Cost of Living units of energy, and when they reach a food item, they gain Prey Value units of energy. If their energy level goes to 0, they die, and if it goes above Energy To Reproduce it splits in half, and gives one-half of the energy to each offspring. The only species considered food items are those listed in Prey Species. If a creature using this procedure lands in a square containing a nonfood item, it tries to replace that item when it leaves so it's not bulldozing its way through the grid, though this doesn't work perfectly.

When a new creature with this action procedure is settled into the world (as opposed to coming from reproduction), it starts out with an energy level equal to one-half of Energy To Reproduce.

Distance to Look
The number of squarest these creatures will look every timestep in search of food.

Speed
The maximum speed, in squares, that a creature of this type can move in one timestep.

Cost of Living
The number of energy units a creature spends each timestep in order to stay alive. This can be a fraction.

Prey Value
The number of energy units gained from each prey item eaten.

Energy to Reproduce
The amount of energy these creatures need to reproduce. When the energy level of a creature reaches this level, it reproduces by splitting in half, with each daughter getting half the energy. New immigrants (which come in with a settlement procedure) start out with an energy level equal to half of Energy to Reproduce.

Prey Species
A list of the species that creatures of this type consider food.

Special Individual Parameters
Each individual stores the species it ate last timestep as its Gut Contents. If it didn't eat anything last timestep, the Gut Contents will be 0. Otherwise, this will be a number corresponding to the species number (count down this many species in the Species Setup window).

Possible Uses
This is the basic move, eat, and reproduce action procedure. I use it for all animal-like species that I want to have eat and reproduce, and where I don't want a fancier procedure. Note that Predator works as well for herbivores and omnivores as it does for carnivores. From the perspective of grass, cows are pretty vicious predators.

Predator In Habitat

This procedure is very much like the Predator procedure above. Creatures look for food, and move towards food with a certain speed. When they eat something, they gain a certain amount of energy, and they use up energy to stay alive. If they reach a certain reproductive energy threshold they reproduce by splitting in half. See the Predator procedure for a more detailed description. The difference between this procedure and Predator is that the creatures here do better in some habitats than others. You can assign a different cost of living to each habitat in the Costs of Living list, and you can also give a probability that a creature will venture into different habitats in the Chance Enter Habitat list.

Costs of Living
A list giving the cost of living for creatures of this species in each different habitat type. The cost of living is how much energy a creature uses every timestep to stay alive.

Chance Enter Habitat
A list of the chance that a creature of this type will enter each type of habitat. Whenever a creature makes a move that would take it from one habitat type to another, it only makes that move with a chance equal to probability given in the Chance Enter Habitat list for the habitat it wants to enter.

Other Parameters
See the parameters of the same names in the Predator procedure.

Possible Uses
Use this when you want a Predator, but you want it to have different parameters in different habitats. For instance, you might want to make some habitats harder to survive in than others. You might also want to make a Predator that wanders around in only one habitat type but is never found in another habitat type.

Random Movement

This procedure moves creatures of this type around randomly at the speed given.

Speed
The speed, in squares per timestep, at which these creatures move. Each movement will be between 0 and Speed squares, independently, in both the horizontal and vertical directions.

Can Move Over
A list of the species these creatures can move on top of.

Possible Uses
I use this when I want items to diffuse through the Grid. For instance, I sometimes use it for food particles in water, or for dispersal of seeds.

Sexual Predator

This works somewhat like the Predator procedure, except that there are two sexes, and in order to reproduce, a member of each sex must meet in the same square.

You must make two species with this action procedure, one for the females and the other for the males. Each creature from both sexes wanders the grid looking for food (defined in Prey Species), which they can see up to Distance to Look squares away, and can move towards at Speed squares per timestep. Each timestep they use Cost of Living units of energy, and each time they land on food they gain Prey Value units of energy. New immigrants and babies (see below) both start out with Initial Energy units of energy; when they reach an energy level of Energy to Reproduce, they start looking for a mate. They are attracted by the species given in Attracted To, which they can see up to Distance to Look squares away.

When a creature of one species lands on the same square as an individual of the Mate Species (which can be the same or different from the Attracted To species), and if this mate has enough energy to reproduce and has not already mated this timestep, they mate and have babies. The two mates have Number of Offspring babies from the mating (this entry should be set the same in both species); these babies represent the species listed in Baby Species, selected randomly. Each mate expends Energy/Offspring energy units per offspring produced (which can be different in the two species). As mentioned above, each offspring starts with Initial Energy units of energy.

Initial Energy
The initial energy level given to new individuals of this species (whether they were settled by a settlement procedure or were produced as babies).

Energy/Offspring
The amount of energy creatures of this species expend for each offspring they produce.

Num Offspring/Mating
The number of offspring produced each time a creature has sex. This should be set the same for the male and female species.

Mate With
A list of the species this species can mate with. Usually, for the male species this list contains the female species, and vice-versa.

Attracted To
A list of species this species is attracted to as potential mates. For most models, this is the same as Mate With. However, you might have other species that mimic the mate species, and you can include those in this list as well.

Baby Species
A list of the possible species which babies of this species can be. Usually this list will contain only the male and female species, giving a 50/50 chance of each baby being male or female.

Other Parameters
See the parameters of the same names in the Predator procedure.

Special Individual Parameters
Each individual stores the energy it will use up for each baby it produces as Energy/Kid, and the timestep at which it last mated in Last Mating.

Possible Uses
This is my generic procedure for species with two sexes. I commonly use it when I want a Predator-like species in which two of them need to come together to mate. If I don't care about actually having two different sexes in the model, I make one species that is hermaphroditic (so that Mate With and Attracted To both contain only the original species, and it mates with itself). I also use the Sexual Predator procedure when I want an Allee effect in a population, since at low population sizes the mates will have trouble finding each other in order to have sex.

Vert Migrate Pred

This procedure is meant to simulate a vertically migrating organism in a marine environment. It is like the Predator procedure, with the added feature of vertical migration. The vertical migration occurs on a given time cycle, the length of which is given in Time Cycle. For instance, if the timestep represents 1 hour and you want the vertical migration to be on a 24-hour cycle, then you would make Time Cycle equal 24. The time at which individuals of this species migrate up is given by Time To Go Up, relative to the time cycle, and the time at which they migrate down is given in Time To Go Down. If you want the individuals to go up at 18:00 every day and back down at 5:00, you would set Time To Go Up to 18 and Time To Go Down to 5. The individuals will migrate down at timestep 5, back up at 18, back down at 29 (24 + 5), back up at 42, and so on. The depth to which the individuals go when they migrate down is given in Depth of Migration. All individuals will go to this depth, then just sit still until it is time for them to migrate up again. When they are sitting still, they expend no energy. When they go back up, they return to the position they were at when they started their descent before starting to feed again.

Time to Go Up
The time at which the creatures start migrating up, relative to the Time Cycle.

Time to Go Down
The time at which the creatures start migrating down, relative to the Time Cycle.

Time Cycle
The number of timesteps in one migration cycle.

Depth of Migration
The row on the grid that the creatures go to when they migrate down.

Other Parameters
See the parameters of the same names in the Predator procedure.

Special Individual Parameters
Each individual stores the number of the row it was in just before it started migrating down as Stored Y Pos, so that it can return there.

Possible Uses
See the Predator Avoidance lab for an example of this procedure.

Visual Predator

This procedure is like the Predator procedure, except that this predator can see different distances in different habitats. The Distance To Look parameter is now a list of numbers, one for each habitat type in the model. In this way, you can simulate predators that are ineffective hunters in some habitats and more effective in others, or have different habitats representing different light levels.
There is one more addition to Visual Predator that is not in Predator. You can limit the vertical range over which the Visual Predator species will move. Top Depth is the highest square that the visual predator will go, and Bottom Depth is the lowest.

Distance To Look
A list of the number of squares creatures of this species can see in each habitat type in the model.

Top Depth
The upper row on the grid above which creatures of this species won't go. Remember that the top grid row is row 1.

Bottom Depth
The lowest row on the grid below which creatures of this species won't go.

Other Parameters
See the parameters of the same names in the Predator procedure.

Possible Uses
See the Predator Avoidance lab for an example of this procedure.

Habitat Procedures

Change on Cycle

This procedure lets you cycle through several habitats as a model runs, or switch from one habitat type to another at a given time. You assign the first habitat this Change on Cycle habitat procedure. The length of the cycle is Time Cycle Length timesteps, and the first habitat changes to the habitat specified in Change To at the timestep given in Change at Time. Change at Time is specified relative to Time Cycle Length. You can also give the second habitat this habitat procedure, and have it change back to the first habitat, thereby cycling back and forth.

You can specify a starting area for a habitat with this procedure using the parameters Left, Right, Top, and Bottom. At time 0, the area specified will be filled with this habitat. If any of these are set to 0, no habitat of this type will be laid down at time 0.

Left
The left side of the habitat if you want a rectangular area of this habitat laid down at time 0. Make this 0 if you don't want any habitat of this type laid down automatically.

Right
The right side of the area of this habitat laid down at time 0.

Top
The top of the area of this habitat laid down at time 0.

Bottom
The bottom of the area of this habitat laid down at time 0.

Time Cycle Length
Determines the length of the cycle on which the changes occur.

Change at Time
The time in the cycle at which habitats of this type are changed into habitats of the Change To type. Change at Time is relative to the cycle length. If Time Cycle Length is 24 and Change at Time is 10, then the habitat will be converted at timestep 10, timestep 34, timestep 58, and so on.

Change To
The type of habitat to change to.

Possible Uses
Suppose you wanted the screen to be yellow for daytime and black for night-time, and your timestep was 1 hour. Make a habitat called daytime, set Time Cycle Length to 24 (hours) and set Change at Time to 20 (20:00, the time that day changes to night). Set Change To to your nighttime habitat. In your night-time habitat, set Time Cycle Length to 24, Change at Time to 6 (6:00, the time that night changes to day), and Change To to your daytime habitat. At timestep 6, all habitat squares that were nighttime would become daytime, and at timestep 20 all habitat squares that were daytime would become nighttime. This cycle would repeat every 24 timesteps, so that night would become day at timestep 30 (24 + 6), day would become night at timestep 44 (24 + 20), night to day at 54 (24 + 24 + 6), and so on.

Periodic Add Rectangle

This procedure periodically adds in a rectangle(s) of the habitat as the model is running. Each rectangle of added habitat is randomly dropped into the Grid, with the constraint that wherever it's dropped the entire rectangle must fit (none of it can extend beyond the edge). The size of each rectangular habitat added is given in Rectangle Width and Rectangle Height. Number Rectangles of these rectangles will be added every ADD ON TIMESTEPS timesteps. If, for instance, Add on Timesteps is 2, rectangles of habitat will be added every other timestep. The rectangles are randomly placed on the grid.

Rectangle Width
The width of the rectangles of habitat that are dropped onto the Grid.

Rectangle Height
The height of the rectangles of habitat that are dropped onto the Grid.

Number Rectangles
The number of rectangles which are dropped onto the Grid in each timestep.

Add on Timesteps
The timesteps at which new habitat is dropped onto the Grid. Rectangles of habitat will be dropped at timesteps that are multiples of Add on Timesteps.

Possible Uses
I've used this procedure to make a fixed rate of habitat destruction over time, in which every year several new parking lots, malls, and housing complexes are built in former forest.

Random Circles

This habitat procedure puts Number circular habitats on the Grid at timestep 0. Each circle of habitat has a radius of Radius, and is randomly placed on the Grid (including locations where part of the circle might extend beyond the edge of the Grid).

Radius
The radius of each habitat circle.

Number
The number of habitat circles to drop onto the Grid.

Possible Uses
As you might expect, I use this procedure to lay down circular habitats. You could use it, for instance, to make a patchy environment.

Regular Circles

This habitat procedure puts in a regularly spaced array of circular habitats, each with the radius of Radius squares, and with a distance between circles given by Dist Between Circles. The circles are laid down at timestep 0.

Radius
The radius of each circle of habitat that is laid down.

Dist Between Circles
The distance between adjacent circles of habitat. The distance between circles is the number of squares from the right edge of one circle to the left edge of the next circle, and from the bottom of a circle to the top of the one below it. This is also the distance from the top and left edges of the Grid to the top-left circle.

Possible Uses
I use this to set up a regular array of patches to look at fragmentation, metapopulations, and other similar phenomena. This is probably the procedure that gets closest to the classic, simple concept of metapopulation habitats.

Regular Circles 2

This habitat procedure makes a regularly spaced array of circular habitats, each with a given radius (Radius), and with a distance from the center of one circle to the center of the next given in Center To Center Dist. The circles are laid down at timestep 0.

Radius
The radius of each circle of habitat that is laid down.

Center To Center Dist
The distance from the center of each circle of habitat to the center of the circles to its left and below it. This is also the distance from the top and left edges of the Grid to the center of the top-left circle.

Possible Uses
See Regular Circles above. The spacing of the circles in this one is a bit easier to explain if you want someone to play with it.

Regular Rectangles

This habitat procedure puts in a regularly spaced array of rectangular habitats, each with the width and height given by Rectangle Width and Rectangle Height. The distances between rectangles are given by Width Between Rects and Height Between Rects. The rectangles are laid down at timestep 0.

Rectangle Width
The width of the rectangles of habitat.

Rectangle Height
The height of the rectangles of habitat.

Width Between Rects
The distance from the right side of one rectangle of habitat to the left side of the next rectangle.

Height Between Rects
The distance from the bottom of one rectangle of habitat to the top of the next rectangle.

Possible Uses
See Regular Circles above.

Regular Rectangles 2

This procedure puts in a regularly spaced array of rectangular habitats, with the size of each rectangle given by Rectangle Width and Rectangle Height, and the spacing of the rectangles given by X Dist From Rect To Rect and Y Dist From Rect to Rect. The habitat is laid down at timestep 0.

Rectangle Width
The width of the rectangles of habitat.

Rectangle Height
The height of the rectangles of habitat.

X Dist From Rect to Rect
The distance from the left side of one rectangle of habitat to the left side of the next rectangle.

Y Dist From Rect to Rect
The distance from the top of one rectangle of habitat to the top of the next rectangle.

Possible Uses
See Regular Circles 2 above.

Single Rectangle

This procedure lays down a single rectangle of habitat at timestep 0. The Left, Top, Right, and Bottom edges of the rectangle are given by the parameters of the same names.

Left
The position of the left side of the habitat.

Top
The position of the top of the habitat.

Right
The position of the right side of the habitat.

Bottom
The position of the bottom of the habitat.

Possible Uses
I've used this to make a single island of habitat, or to divide the Grid into two different sections, or for a number of other situations in which I want one swath of the Grid different from another.

Sampling Techniques

There are several parameters common to all sampling techniques, and these are described in the section on sampling. To set which species will get sampled, bring up the Sampling Setup box and select the SPECIES TO SAMPLE button. To set where the samples are to be saved, click on the SET SAVE FILE button. To set advanced features of sampling, such as automatic sampling and the verbosity, click the ADVANCED FEATURES button.

Destruct Placed Quadrat

This technique lets you use the mouse to place a quadrat on the grid. It then samples the quadrat and kills everything it samples. You specify the width and height of the quadrat you want to use in Quadrat Width and Quadrat Height. When you take a sample, a rectangle will appear on the grid. This rectangle is controlled by the mouse. Use the mouse to position it, then click on the mouse button to take a sample. Any creatures of the Species to Sample inside this rectangle have a chance of being sampled and killed. The chance per species of being sampled and killed is specified in Percent Caught. If you give a probability of 0.5 for a species to be sampled, then on average 50% of the creatures of that species that fall inside the quadrat will be seen by the sampling technique. The technique kills those creatures that it sees and reports the number it killed. It only reports those creatures it kills—the other 50% get away—they are neither sampled nor killed.

Quadrat Width
The width of the sampling quadrats.

Quadrat Height
The height of the sampling quadrats.

Percent Caught
The percent of the creatures within each quadrat that are sampled.

Possible Uses
I use this technique to simulate a fishing net.

Line Transect

This procedure samples a series of either horizontal or vertical lines on the grid. The number of sampling lines to use is specified by Number of Transects. Each transect is randomly placed, and the user is asked if they want to sample that line. If the answer is NO, a new random location is picked. If the answer is YES, the sampling technique reports how the number of each sampled species under the line.

In addition, this technique includes a method for sampling the sizes of creatures that can extend for more than one square. EcoBeaker is not really built for creatures to be larger than one square, but there are one or two Individualistic action procedures that will build multisquare creatures. In these creatures, one of the individualistic parameters will specify which individual each grid square belongs to. In the Large Species list you tell the transect sampler which species can be larger than one square. Then you must tell it which individualistic parameter specifies the individual number of each creature, by setting Example Large Species and Creature Num. Now when you sample, the Line Transect technique will report the sizes of each individual creature that it encounters, with a summary at the end giving the total number of squares covered by each species.

Note: First, all large species must use the same Individualistic action procedure to make them large; otherwise, the individualistic parameter specifying the creature number may not be the same for different species. Second, if you don't have any large species, simply uncheck every species in the Large Species list (the default) and ignore all of this.

Number of Transects
The number of transects to use each time the user requests a sample.

Horiz ('0') or Vert ('1')
If this is set to 0, transects will be placed horizontally across the Grid, otherwise they'll be placed vertically. In either case, the transects will go from one side of the Grid to the other.

Large Species
The list of species in which an individual can extend for more than one square. If you don't have any creatures larger than one square, uncheck all species in this list and ignore the next two parameters.

Example Large Species
If you have large species, this should be set to one of them.

Creature Num
If you have large species, select one of these species in Example Large Species. Creature Num will then contain a list of all the individualistic parameters of the Example Large Species. From this list, select the parameter that gives the number of individual creatures (should be labeled something like Creature Number or Indiv Num).

Possible Uses
I use this to simulate line transects, particularly for situations in which creatures may exceed one square and I want to know individual sizes.

Mark Recapture

This technique puts out an array of traps, with Num Traps Horiz traps across the Grid and Num Traps Vert traps from top to bottom of the Grid. Each trap has a width and height of Trap Size. The traps are left on the grid until Number to Capture creatures wander into them. When a creature is caught, the user is asked if they wish to mark it, and, assuming they choose YES, the creature is marked and then released. Marking occurs by changing the creature into another species, called the Marker Species, which you should set up to be the same as the species you're sampling, but select a different color. When Number to Capture creatures have been caught and marked, the trap disappears for Time To Wait timesteps, then reappears and starts catching both the original and the marker species. Each time it catches an individual from either of these species, it is reported to the user. When it has caught a combined total of unmarked and marked creatures equal to Number to Capture, a summary is displayed, and the trap disappears.

A few notes on this technique. First, you can only sample one species at a time, so only the first species given in the Species To Sample list is sampled. Second, if for some reason you save a situation file while you are mark-recapture sampling, or reset before sampling is finished, you will probably get Automatic Sampling enabled with some odd parameters, producing confusion next time you use that sampling procedure. When using mark-recapture, SAVE and RESET before or after you try sampling. Finally, you can use this technique only to sample individualistic species, not Grid-Based species.

Trap Size
The height and width of each trap.

Num Traps Horiz
The number of columns of traps.

Num Traps Vert
The number of rows of traps. The total number of traps is Num Traps Horiz times Num Traps Vert.

Number to Capture
The number of creatures to capture and mark; also the number of creatures to recapture (both marked and unmarked).

Time to Wait
The number of timesteps to wait between the end of capturing and the beginning of recapturing.

Marker Species
When a creature is captured, it is marked by turning it into Marker Species. The marker species should have the same action procedure as the original species, but it should be a different color.

Possible Uses
The obvious, to simulate mark-recapture sampling. You can be fancier and simulate marking having a particular effect on creatures (such as increasing their death rates) by changing the parameters of the marker species relative to the original species (but keep the same action procedure).

Percent Cover

This procedure samples a series of either horizontal or vertical lines on the Grid (specified in Horiz ('0') or Vert ('1'). The number of lines to use is specified by Number of Transects. Each transect is randomly placed, and the user is asked if they wish to sample that transect or not. If they choose NO, a new random location is picked. When they choose YES, the sampling technique determines the percent cover of each of the sampled species underneath the line and reports this to the user. It then reports a combined percent cover for all of the lines at the end.

Number of Transects
The number of transects to use each time the user requests a sample.

Horiz ('0') or Vert ('1')
If this is set to 0, transects will be placed horizontally across the Grid, otherwise they'll be placed vertically. In either case, the transects will go from one side of the Grid to the other.

Possible Uses

This is identical to the Line Transect technique, but results are reported in percentages instead of number of squares covered (and it doesn't have the large-species feature).

Placed Quadrat

This sampling technique lets you place a quadrat on the Grid using the mouse. You specify the width and height of the quadrat in Quadrat Width and Quadrat Height. When you take a sample, a rectangle will appear on the Grid, which is controlled by the mouse. When you click the mouse button, the area under the rectangle will be sampled. You can also set a maximum total number of samples taken in Total Samples Allowed. If you only want the user to be able to take 100 samples, then set Total Samples Allowed to 100. This is the total number of samples that the user could take for the entire time the model is running, even if the user resets the model.

Quadrat Width

The width of the sampling quadrats, in Grid squares.

Quadrat Height

The height of the sampling quadrats, in Grid squares.

Total Samples Allowed

The total number of quadrats the user is allowed to take. The user will be limited to this total even if they reset the model, so short of quitting and reloading the model there is no way to get around this limit. If you don't want to limit the number of samples, set this to a much larger number.

Possible Uses

I used this feature when designing a sampling lab in which I wanted to allow only a certain amount of sampling effort, and didn't want users to be able to cheat by resetting the model and trying again. I also use it in other situations in which I want to be able to manually place the sampling quadrats.

Random Quadrat

This sampling technique throws a number of quadrats out in random locations within the Grid, and counts what's inside of each one. The number of quadrats to sample is specified by Number of Quadrats, and each quadrat is Quadrat Width wide by Quadrat Height tall. Each time a quadrat is put down, the user is asked if they wish to sample in that location. If the answer is NO, the quadrat is moved to a different random location. When the answer is YES, a dialog box appears with the contents of the quadrat.

Quadrat Width
The width of the sampling quadrats, in Grid squares.

Quadrat Height
The height of the sampling quadrats, in Grid squares.

Number of Quadrats
The number of quadrats to use each time the user elects to sample.

Possible Uses
Use this wherever randomly placed quadrats might be an appropriate sampling technique (and even in some places, such as with moving animals, where quadrat sampling might be harder in real life than it is in EcoBeaker).

Systematic

This sampling technique samples a series of quadrats that are arranged systematically in a grid pattern. Each quadrat is Quadrat Width wide by Quadrat Height tall, the horizontal distance between the end of one quadrat and the beginning of the next is Horiz Dist Between Quads, and the vertical distance between quadrats is Vert Dist Between Quads. The quadrats are sampled from top to bottom and left to right. When the user takes a sample, the sampler moves through each quadrat and asks if the user wants to sample it. If the answer is YES, the number of creatures from each sampled species within that quadrat is displayed. A total is displayed at the end.

Quadrat Width
The width of the sampling quadrats, in Grid squares.

Quadrat Height
The height of the sampling quadrats, in Grid squares.

Horiz Dist Between Quads
The distance from the right edge of one sampling quadrat to the left edge of the next one, in Grid squares.

Vert Dist Between Quads
The distance from the bottom edge of one sampling quadrat to the top edge of the next one, in Grid squares.

Possible Uses
I included this as a contrast to the Random Quadrat technique, so users can compare sampling an area systematically with randomly sampling the same area.

T-Square

The T-Square sampling technique is a nearest-neighbor style of sampling technique. The idea is that it randomly lays down a series of points, and for each point, it measures the distance from that point to the nearest individual. Then the sampler starts from that individual, and looks for its nearest neighbor. However, it doesn't use any individuals that are on the same side of the individual as the random point was. In other words, if you draw a line from the random point to the individual, and then another line from the individual to its nearest neighbor, the angle between the two lines must be greater than 90 degrees. From these two distances, you can get a measurement of the aggregation of the species; if the species is randomly distributed, you can estimate the population size. Look in an ecological statistics book for the formulas.

Number of Points
The number of sampling points to use.

Possible Uses
I've used this only in a lab investigating nearest-neighbor sampling techniques and measuring patchiness.

Statistics

Autocorrelation

This statistic calculates Moran's I at a series of lags that you specify, either over all the rows or all the columns of the grid (type in '0' for rows, or '1' for columns). This is done by calculating the autocorrelation for each row at each lag, then averaging across the rows (or columns). There are, of course, other ways that you might want to calculate autocorrelation (on a single row or column, in all directions at once, and so on), and as you request these, I will add them.

Note that you can only use this statistic in histograms, and you set the lags for which autocorrelation is displayed in the Scales section of the histogram, not in the statistic setup box. For more information, refer to the section on histograms.

Rows ('0') or Columns ('1')
Put a 0 here to get average autocorrelation along the rows of the Grid, and a 1 to get average autocorrelation along the columns of the Grid.

Possible Uses
Use to get a measure of autocorrelation, which you can relate to lots of other measures. Note that this statistic can be very slow, since it takes quite a bit of math to calculate the autocorrelation across a whole Grid, especially if the Grid is large.

Avg Individual Param

For models in EcoBeaker that use Individualistic species, in which each creature has individual parameters associated with it, you may want to know the value of one of the individual parameters. You can get this value for a specific individual by clicking on it (see the "Getting Info About Creatures" section of this manual) or you can get an average of the parameter over all the individuals using this statistic. Select the Species for which you wish to see an individual parameter, then select the individual parameter that you want from that species in Parameter.

Species
The species for which you want to measure an individual parameter.

Parameter
The individual parameter that you wish to measure.

Possible Uses
I've found this particularly useful in evolution models in which I want to watch a particular parameter as it evolves, or another parameter, such as energy level, that describes how well the creatures of a particular species are faring.

Note: If, instead of an average, you want to look at a histogram of an individual parameter, use the Indiv Param Histo statistic. If you want to know the sum of the individual parameter across all individuals, use the Total Individual Param statistic. Note also that this statistic has no meaning for Grid-Based species.

Avg Number Neighbors

This statistic goes through each individual of the Species you specify, and calculates how many creatures of the same or another species specified in Neighbor are within Neighborhood Distance of the individual. The statistic then reports this average for all individuals.

Species
The focal species whose neighbors you want to count.

Neighbor
The species of the neighbors that you're interested in. To look at intraspecific crowding, select the same species in Species and Neighbor.

Neighborhood Distance
The radius of the neighborhood within which you want to count neighbors.

Possible Uses
The number of neighbors within a certain radius of each individual is an indication of whether or not the individuals are aggregated. You can also look for signs of interspecies aggregation by looking at how many individuals of another species are near each individual of the first species. For a given density, the more individuals there are close by to other individuals, the more aggregated those individuals are.

Habitat Area

This statistic measures the number of squares of the Grid which are covered by the habitat given in Habitat.

Habitat
The habitat type that you wish to measure.

Possible Uses
Use this to find out how much of the Grid is covered by a given habitat. This is primarily useful for habitats that are dropped down in random places (and so could cover up pieces of each other) or are periodically added to.

Indiv Param Histo

You may want to know the distribution of an individual parameter for creatures of a particular species. This statistic makes a histogram of an individual parameter from an Individualistic species. Select an individualistic Species, then select, in Parameter, a parameter of that species for which you want a histogram.

Note that you can use this statistic only in a histogram, not any other type of graph, and the bins for the histogram are set in the Scales section of the histogram, not in the statistic setup box. For more information, refer to the section on histograms.

Species
The species for which you want a histogram of an individual parameter.

Parameter
The individual parameter of this species that you want to see.

Possible Uses
Use this measurement when the distribution of a particular individual parameter, not just its average, is important. If you want only the average of an individual parameter, use the Avg Individual Param statistic.

Lloyd's Index

Lloyd's index is one of several statistics that use the ratio of variance to mean in population sizes in different sampled areas to say something about how aggregated a species is. The formula for Lloyd's index is:

$$Lloyd's = \frac{mean + \left(\frac{variance}{mean} - 1 \right)}{mean}$$

where mean is the mean population size in each area sampled, and variance is the variance in population sizes between the areas sampled. Lloyd's index needs densities, and EcoBeaker models typically have only one creature per square. Therefore, to use Lloyd's index, we need to group squares together to get densities. The number of squares to group together is given by Quadrat Length, and the species for which to calculate Lloyds is given in Species. The statistic will move through the grid and group the squares together into blocks that are Quadrat Length on each side, determine the density of the Species within each of these blocks, then calculate Lloyd's index based on that data.

Species
The species for which to calculate Lloyd's index.

Quadrat Length
The width and height of the quadrats used to calculate the index. The statistic will calculate the density of creatures within each quadrat. Quadrats are placed one next to the other. Squares on the bottom and right sides of the Grid that don't fit a whole quadrat will not be used in the calculation.

Possible Uses
I use this as a simple determinant of patchiness. In general, I don't think it's very useful, but it is an example of this class of nondimensional patchiness statistics.

Percent Cover

This statistic gives the percent of the grid covered by a species. Set Species to the species for which you want the percent cover. Percent cover is the number of squares occupied by a creature of that species divided by the total number of squares in the grid.

Species
The species for which you want to know percent cover.

Possible Uses
Use to find the percent of the Grid covered by some species.

Sampled Pop Size

Most sampling techniques will calculate an estimated population size for each species sampled, based on the last sample taken. The sampling techniques do not report this estimate at the time of sampling, but you can see it by using this statistic. Set Species to the species for which you want an estimated population size. Each time you take a sample, this statistic will be updated.

Species
The species for which you want to know the estimated population size.

Possible Uses
This statistic is useful in comparing sampled population sizes to actual population sizes. Note that there are some sampling techniques that don't make an estimated population size.

Shannon's Diversity Index

This calculates Shannon's index of species diversity. The formula is:

Shannon's diversity index $H = -\sum_{i=1}^{S} p_i \ln(p_i)$

where S is the number of species and p_i is the population size of species i divided by the total population size of all species (the percentage of individuals which are of species i). Look in an ecological statistics text for more information.

Include Empty (No = '0', Yes = '1')
If you want empty squares to be included in the calculation of the diversity index, put 1 for this parameter, otherwise put 0.

Possible Uses
I use this index or Simpson's diversity index as a simple measure of species diversity.

Simpson's Diversity Index

This calculates Simpson's index of species diversity. The formula is:

Simpson's diversity index $D = \dfrac{1}{\sum\limits_{i=1}^{S} p_i^2}$

where S is the number of species and p_i is the population size of species i divided by the total population size of all species (the percentage of individuals which are of species i). Look in an ecological statistics text for more information.

Include Empty (No = '0', Yes = '1')
If you want empty squares to be included in the calculation of the diversity index, put 1 for this parameter, otherwise put 0.

Possible Uses
I use this index or Shannon's diversity index as a simple measure of species diversity.

Species Population

This statistic gives the population size of the species specified in Species.

Species
The species for which you want to know population size.

Possible Uses
This is the statistic I use the most, since much of ecology is concerned with population sizes. Many times I include a graph with population sizes even in models where it's not directly relevant, just to have an idea of what's going on.

Spec Pop in Area

This statistic gives the population size of the species specified in Species in the area on the Grid specified by Area Left, Area Top, Area Right, and Area Bottom.

Species
The species for which you want to know population size.

Area Left
The left side of the area in which you're measuring population size.

Area Top
The top side of the area in which you're measuring population size.

Area Right
The right side of the area in which you're measuring population size. Note that the area goes up to, but does not include, the squares in the Area Right column.

Area Bottom
The bottom side of the area in which you're measuring population size. Note that the area goes up to, but does not include, the squares in the Area Bottom row.

Possible Uses
I use this when I want to know the population size of a species only within a certain area, for instance on an island or a patch of habitat.

Total Individual Param

For models in EcoBeaker that use Individualistic species, in which each creature has individual parameters associated with it, you may want to know the total value of one of the individual parameters, summed across all individuals. Select the Species for which you want to see an individual parameter, then select, in Parameter, the individual parameter you want from that species. This statistic has no meaning for species that are not Individualistic.

Species
The species for which you want a sum of an individual parameter.

Parameter
The individual parameter of this species for which you want a sum.

Possible Uses
This statistic is similar to Avg Individual Param, but gives the sum of the parameter across all individuals, instead of the average. You will use one or the other of these statistics depending on what the parameter represents and exactly how you are trying to interpret the values. You may also want to see the distribution of the individual parameter, in which case you can use the Avg Individual Param statistic.

User Programming

You may want to try making your own procedures such as the ones described above. I am making available all the source code you need to do this. In addition, on the disk there is a programmer's manual to guide you in writing EcoBeaker procedures. I have tried to make it as easy as possible to add new procedures, and I can add a simple procedure in ten minutes, including all the user interface, graphics, and so on. I am hoping that the learning curve for others to do this is not too steep, but so far I am the only one who has written any procedures, so I can't tell. If you are interested, see the "Installation" section of this manual to install the extra material you'll need for programming EcoBeaker. You'll also need to download some precompiled code, as described in that section. Finally, you'll need to buy a copy of Symantec C++ for Macintosh or PowerPC, since EcoBeaker is written using their class library. You can also use CodeWarrior to do the actual compiling—my PowerPC version used to be compiled with CodeWarrior. However, you'll still need to buy Symnatec to get their class library.

If you try to program your own procedures in EcoBeaker, I would be very interested in hearing how it went, whether you were successful, and what I could do to make the process easier. I'm also happy to answer your questions.

Glossary

Action procedure

A set of rules that controls the actions of all members of a species. There are many different Action procedures in EcoBeaker; you choose one for each species in a model. Each Action procedure has a set of associated parameters that further control its action. These are described in the "Grid-Based Action Procedures" and "Individualistic Action Procedures" sections of this manual. Also, see the "Species" section of the manual for further explanation.

Creature

EcoBeaker models are written around Species. I call each individual of a certain species a "creature." Think of a creature as one plant, one animal, one hamburger, or something along those lines.

Grid

There are two Grids in the program, the Species Grid and the Habitat Grid. The Species Grid is the two-dimensional area in which all the action takes place. This is where the creatures run around. You can think of it as the world within which models take place; it appears in the Species Grid window on the screen. This is generally what I'm referring to when I say Grid. The Habitat Grid is an accompanying area that stores Habitats, if you have Habitats in your model. See the "Setting up the Grid" section of this manual for further explanation.

Grid-Based vs. Individualistic Species/Creatures

Grid-Based creatures are stored as a colored square on the screen, and nowhere else (thus they are based on the Grid). Within species each Grid-Based creature is identical. Individualistic creatures are shown on the grid, but their position is also stored separately in a big array. Along with the position of each individual, you can store other information (e.g., energy level, direction it's facing, and so forth). Thus each of these creatures is an individual, distinctly different from other creatures of the same species (though they follow the same set of rules). See the "Species" section of this manual for further explanation.

Grid square

Each position in the Grid is represented by a square on the screen; thus I call each grid position a "grid square," or "square," for short.

Habitat

A type of terrain in a model. Each habitat has a set of rules associated with it, called a Habitat procedure, which determines the habitat's distribution across the Grid. See the "Habitats" section of this manual for further explanation.

Habitat procedure

A set of rules determining the distribution of a habitat across the Grid. Descriptions of each Habitat procedure are given in the "Habitat Procedures" section of this manual.

Hidden Creatures

When you have only Grid-Based creatures in your model, each square on the Grid can hold only a single creature. This is due to the fact that if another creature moves into the square, the original creature's color is replaced by the new creature's color, and the old creature is lost. However, with Individualistic creatures the position of each creature is stored independently of the Grid, so more than one creature can occupy the same square. Still, only one creature can be plotted in a square. This means that other Individualistic creatures may be hidden under the top one. Looking for these hidden creatures is somewhat expensive computationally, so some procedures only look for visible creatures, while others give you a choice of whether or not to look for hidden ones as well.

Individual

The same as a creature—a single organism. Each individual belongs to one of the species in the model.

Individual parameter

Each creature from an Individualistic species has a set of parameters associated with it, and these are called Individual Parameters. The first two of these parameters will always be the (x,y) position of the creature, and any further parameters will depend on the Individual action procedure being used.

Individualistic action procedure

A subset of the Action procedures that are used for Individualistic species.

Individualistic species/creature

See Grid-Based vs. Individualistic Species above.

Parameter

All procedures, statistics, and sampling techniques in EcoBeaker have a set of parameters that control their function. These parameters are settings the user can change to modify the way the procedure works. Explanations of the parameters for each procedure are given under the description of the procedure in this manual.

Sampling technique

Sampling techniques are procedures that let you sample the distribution of species in the Grid as you would in real life. They give answers with error, as opposed to Statistics, which give completely accurate answers. Descriptions of each Sampling Technique are given in the "Sampling Techniques" section of this manual.

Settlement procedure

A set of rules that controls how new individuals of a species enter the Grid from the outside world. Each Settlement procedure has a set of associated parameters that further control what it does. These are all described in the "Settlement Procedures" section of this manual. Also see the "Species" section of the manual for further explanation.

Species

The basic modeling unit in EcoBeaker. A Species is a set of rules governing how members of that species get onto the Grid and what they do once they are on the Grid.

Statistic

A procedure that calculates something that can be plotted in a graph. Descriptions of each statistic are given in the "Statistics" section of this manual.

Timestep

The models in EcoBeaker are run in discrete time, and a timestep is one unit of modeling time. Everything in a model happens once per timestep. Depending on your model, a timestep may be anywhere from a second to a year or more.

Transition Matrix

For each Species, transition matrix gives the chance, per timestep, that a creature of that species will transform into a creature of another species. See the "Transition Matrix" section of this manual for further description.

Notes